Control of Poultry Mites (*Dermanyssus*)

Olivier A.E. Sparagano
Editor

Control of Poultry Mites (*Dermanyssus*)

Previously published in *Experimental and Applied Acarology* Volume 48, Nos. 1–2, 2009

Springer

Editor
Olivier A.E. Sparagano
Newcastle University
School of Agriculture,
Food and Rural Development
Newcastle upon Tyne
NE1 7RU, UK
olivier.sparagano@ncl.ac.uk

ISBN: 978-90-481-2730-6 e-ISBN: 978-90-481-2731-3

DOI: 10.1007/978-90-481-2731-3

Library of Congress Control Number: 2009927096

Cover Illustrations:

Left: European chicken farm under heavy red mite infestation. Picture by Olivier Sparagano.

Right: Female Hypoaspis aculeifer attacking Dermanyssus gallinae. Picture by Izabela Lesna, Urs Wyss and Maurice W. Sabelis.

Printed on acid-free paper

Springer.com

Contents

Control of poultry mites: where do we stand?

Olivier Sparagano

Originally published in the journal Experimental and Applied Acarology, Volume 48, Nos 1–2, 1–2.
DOI: 10.1007/s10493-009-9259-x

This special issue of Experimental and Applied Acarology is critically arriving at the right time. Since *Salmonella* vaccination is now widespread within the poultry industry, it seems the new economic, welfare and epidemiological problem is now the poultry red mite, *Dermanyssus gallinae* (also called red poultry mite, poultry mite, red mite or chicken mite). In 2012 the European Commission will ban traditional cage systems to move towards enriched cages or more open systems, such as free range or barns, which unfortunately could see mite population rocketing as they would be able to hide and proliferate better under these open environments.

Like with many blood-feeding arthropods the consequences of their attacks are multiple: welfare issues of birds, becoming anaemic, picking feather, becoming restless and aggressing each other; egg production going down, because of the increased fragility of the egg shell; blood staining of the eggs, usually leading to refusal by supermarket corporations; and the mites are reservoirs of—and transmit—several bacterial and viral poultry diseases, which may also have nasty effects on man.

So the eradication or at least the control of this ectoparasite would greatly impact several aspects of our communities. However, acaricide products used to try controlling the poultry red mite have shown some limitations, either because mites became more and more resistant (Marangi, Cafiero et al., and Roy, Chauve et al., both in this issue), or because some products are withdrawn from the national market because of negative impacts on the environment as a side effect. The costs for prevention and control are globally impressive (Sparagano et al., this issue) and therefore control strategies are needed.

The 14 papers published in this special issue (prepared by 43 co-authors from 11 countries) are highlighting the state-of-art of our knowledge about these ectoparasites, but also about research initiatives in laboratories and on farms to stop this proliferating parasite. Colleagues present papers about physical control methods (Kilpinen and Steenberg), or biological methods based on plants (George et al.), the use of predators (Lesna et al.), an immunological approach (Wright et al.), or a mixture of them (Maurer et al.). Owing to the

O. Sparagano (✉)
School of Agriculture, Food and Rural Development, Newcastle University, Newcastle upon Tyne NE1 7RU, UK
e-mail: Olivier.sparagano@ncl.ac.uk

recent developments in molecular genetics we now understand better the variability between mite populations at farm or country levels (Marangi, De Luna et al., Roy, Dowling et al.), or what pathogens (Valiente-Moro et al.) or symbionts (De Luna et al.) are associated with this poultry red mite, which could become either a pathogen reservoir or a way to knock down the mite by eliminating its symbiotic population.

Finally, colleagues present practical information about monitoring and prevention of this devastating pest for humans and animals (Mul and Koenraadt; Zenner et al.). This special issue should give the reader not only a broad overview of the field, but also stimulate ideas for further collaborations, networking and investigations for many years to come.

Prevalence and key figures for the poultry red mite *Dermanyssus gallinae* infections in poultry farm systems

Olivier Sparagano · Aleksandar Pavlićević · Takako Murano · Antonio Camarda · Hamid Sahibi · Ole Kilpinen · Monique Mul · Rick van Emous · Sophie le Bouquin · Kristian Hoel · Maria Assunta Cafiero

Originally published in the journal Experimental and Applied Acarology, Volume 48, Nos 1–2, 3–10.
DOI: 10.1007/s10493-008-9233-z

Abstract Recent surveys and sample collection have confirmed the endemicity of *Dermanyssus gallinae* in poultry farming worldwide. The reduction in number and efficacy of many acaricide products has accentuated the prevalence rates of this poultry ectoparasite observed more often in non intensive systems such as free-range, barns or backyards and more often in laying hens than in broiler birds. The lack of knowledge from producers and the utilisation of inadequate, ineffective or illegal chemicals in many countries have been responsible for the increase in infestation rates due to the spread of acaricide resistance. The costs for control methods and treatment are showing the tremendous economic impact of this ectoparasite on poultry meat and egg industries. This paper reviews the prevalence

O. Sparagano
School of Agriculture, Food and Rural Development, Newcastle University, Agriculture Building, Newcastle Upon Tyne NE1 7RU, UK

A. Pavlićević
AVES Inc, Palic, Serbia

T. Murano
Chiba Prefectural Livestock Research Center, He 16-1 Yachimata, Yachimata, Chiba 289-1113, Japan

A. Camarda
Dipartimento di Sanità Pubblica e Zootecnia, Facoltà di Medicina Veterinaria, Università degli Studi di Bari, 70100 Valenzano, Italy

H. Sahibi
Département de Pathologie et de Santé Publique Vétérinaires, Unité de Parasitologie et Maladies Parasitaires, Institut Agronomique et Vétérinaire Hassan II, BP 6202, Rabat-Institut, Rabat, Morocco

O. Kilpinen
Institute of Integrated Pest Management, Danish Pest Infestation Laboratory, University of Aarhus, Skovbrynet 14, 2800 Kgs. Lyngby, Denmark

M. Mul · R. van Emous
Animal Sciences Group of Wageningen UR, P.O. Box 65, 8200 AB Lelystad, The Netherlands

rates of this poultry pest in different countries and for different farming systems and the production parameters which could be linked to this pest proliferation.

Keywords *Dermanyssus gallinae* · Prevalence · Infestation rates · Control costs · Poultry

Introduction

Dermanyssus gallinae (De Geer, 1778) also known as the poultry red mite (PRM) or the poultry mite is an increasing epidemiological and economical problem for the poultry industry worldwide. This ectoparasite is a blood feeder and is responsible for egg downgrading and spotting, anaemia in birds and more reports suggest it could have a vector role for several human and animal diseases. It is the most important ectoparasite affecting laying hens (Chauve 1998).

The current European legislation which will ban by 2012 traditional cages for poultry birds (European Council Directive 1999/74/EC) and the removal of acaricide products from national markets due to the increase in acaricide resistance or welfare concerns will have a tremendous impact on the proliferation of such pest which has shown in this paper in endemic in many countries and is becoming the most serious deleterious ectoparasite in poultry farming systems worldwide. New control methods highlighted in other papers within this special *Dermanyssus* issue in Experimental and Applied Acarology, show the need to urgently tackle such parasites to reduce economical losses, improve welfare, and control zoonotic risks for farming workers.

Poultry data

Poultry production is an important and increasing meat/egg market with millions of birds grown in participating countries as shown on Table 1 in parallel of the staggering 8.56 billion birds raised annually in the USA.

Caged animals are making the most of the laying hen production systems up to 100% in Japan for instance with free range and barns systems making the rest, in some countries, while the organic production system (although increasing in some developed countries) represent only a few percents of the global market.

The introduction of "enriched cages" in some countries could become a major production possibility if producers cannot convert their traditional cage systems into less extensive systems. However, the use of enriched cages with nesting boxes could help the mites to better survive, hide, and therefore infest more poultry in this new system supposedly improving birds' welfare. By improving animal welfare enriched cages could unfortunately also optimise survival conditions for the poultry red mites (Chirico and Tausan 2002).

S. le Bouquin
Unité EBEAC, AFSSA, BP 53, 22 440 Ploufragan, France

K. Hoel
Animalia, P.O. Box 396, Økern, 0513 Oslo, Norway

M. A. Cafiero (✉)
Istituto Zooprofilattico Sperimentale della Puglia e della Basilicata, Via Manfredonia, 20, 71100 Foggia, Italy
e-mail: ma.cafiero@izsfg.it

Table 1 Key data for poultry production and *Dermanyssus gallinae* prevalence

Country	Annual poultry production in million birds (average flock)	% in traditional cages	% in enriched cages	% in barns	% in free-range	% in organic systems	% in backyards	Other systems	*Dermanyssus* prevalence[a] (%)	Estimated annual cost of *Dermanyssus*
Denmark	2.7 (11,700)	56	<1	23	6	15	Unknown	Unknown	C: 32 B: 50 FR: 68 Organic: 36	Unknown
France	46.5 for laying hens and 111 for broilers (cages: 39,800; other systems: 5,700)	76.5	4.6	3.4	8.6	3.0	Unknown	8% "Red Label"	C: 72 B: 50 FR: 56 Organic: 80	Cages: 4.33 €/100 birds; alternative systems 3.83 €/100 birds)
Italy	486 including 435 for broilers and 51 for layers (15,000–20,000)	96.4	Unknown	2.4	0.5	0.7	Unknown		C: 74.1	Unknown
Japan	860 (unknown)	Circa 100	0	<1.0	0	<1.0	<1.0		C Layers: 85.2 C for broilers: 0.6	66.85 million €
Montenegro	0.43 (2,500–25,000)	87	4.0	3.75	1.00	Unknown	3.75	None	C layers 30–80	Unknown
Morocco	294 (unknown)	Unknown	Unknown	Unknown	Unknown	Unknown	Unknown	Unknown	BY: 90 C broilers: 20 C layers: 55	Unknown
Norway	3.6 (1,900)	54.0	26.0	18.0	0	2.0	0	None	C layers: 23	Unknown
Serbia	80.0 (unknown)	Unknown	Unknown	Unknown	Unknown	Unknown	Unknown	Unknown	C layers: 90	Unknown
The Netherlands	30.12 (26,600)	46.0	2.0	40	12.0	2.0	None	None	C: 82 B: 83 Organic: 78	11.0 million €
UK	860 (10,380)	60.0	Unknown	4.0	30.0	6.0	Unknown	Unknown	C: 7.5–87.5 B: 32.5 FR: 60.0	3 million €

[a] *C* cages, *B* barns, *FR* free-range, *BY* backyard

Red mite prevalence

Infestation rates can reach 80–90% of poultry birds as observed in the United Kingdom (UK), Italy, Serbia, Morocco, Japan, Montenegro, and The Netherlands (Table 1). Less intensive farming systems such as barns, free range and organic farming are often showing higher prevalence rates due to the greater potential for *D. gallinae* to hide in cracks and crevices and avoid chemical control methods. For instance, Höglund et al. (1995) observed only 6% infestations in cage systems but 33% in alternative systems and 67% of backyard flocks being infested. Similar figures were observed in UK with 7.5, 32.5 and 60% for the above three poultry systems, respectively (Anon 2003). However, as shown in Table 1 there is no prevalence trends between poultry systems as different countries show different prevalence rates. In The Netherlands, poultry husbandry advisers estimate a prevalence of 95%. Considering that many countries will ban cages there is therefore a risk that *D. gallinae* prevalence will increase with higher economical losses for the farming industry if such pest is not under control rapidly.

Associated costs

The cost of *D. gallinae* is difficult to evaluate on a global scale but some colleagues have been able to calculate costs at national levels such as 4.33 €/100 birds and 3.83 €/100 birds for cage and alternative systems in France, respectively (Lubac et al. 2003). In The Netherlands, Dutch poultry farmers estimated the costs for preventive and control measures to be €0.14 per hen per round and de damage due to RPM because of higher feed intake, higher mortality, and lower egg quality were estimated as €0.29 per hen per round (Emous et al. 2005) or as shown on Table 1 representing millions of euros/dollars in production and animal losses, treatment, veterinary bills, and lost working days.

Human costs are difficult to establish but cases of dermatitis related to *D. gallinae* are now more and more obvious while workers in some countries had to be paid 3 times more in recent years to work with *D. gallinae* infested birds (Sahibi et al. 2008). In Egypt, a report showed similar attacks on farm workers from *Ornithonyssus* mites (Mazyad and Abel El-Kadi 2005) while it was with *D. gallinae* in Israeli poultry workers (Rosen et al. 2002). As *D. gallinae* is also feeding on synantropic birds, including pigeons and sparrows, more cases have recently been published on human attacks due to bird nests found in close proximity to private households (Rosen et al. 2002; Cafiero et al. 2008), hospitals (Sexton and Barton 1975; Auger et al. 1979; Regan et al. 1987; Bellanger et al. 2008) or offices (Cafiero et al. 2007). Furthermore, the red mite can feed also on the wild birds (Kristofik et al. 1996) or on other animals (Ramsay et al. 1975; DeClercq and Nachtegaele 1993; Mignon and Losson 2008).

Prevalence seems to be dependent of several parameters

The research done for this paper showed that in Southern Italy (Apulia region) farm sizes had an important impact on prevalence with small farms (1,000–5,000 birds) showing a prevalence of 92.3% while bigger farms (5,000–20,000 birds) showed a prevalence of 55.9% only (in laying birds). These results are higher than the 20% infestation rate previously observed in the Italian Abruzzo region (central region of Italy) in traditional (free-range) poultry farms (Paoletti et al. 2006).

Change of flocks and repopulation can have a tremendous impact on *D. gallinae* as observed in Montenegro where the prevalence in layers in cages was 30% at the beginning of 2007 but dramatically rose to 80% after flocks were repopulated.

Fig. 1 European farm heavily infested with *Dermanyssus gallinae*

In France, reports show an endemic situation with almost all part of the country showing infestations (Beugnet et al. 1997; Chauve 1998) with a higher prevalence during the winter (Lubac et al. 2003) whereas in Denmark worst infestation cases are observed in the late summer (personal communication, Kilpinen, Lyngby, Denmark) and also in Italy (personal communication, Camarda, University of Bari, Italy).

In UK, several authors have also reported high prevalence rates (Guy et al. 2004; Fiddes et al. 2005).

Bad hygiene practices will have dramatic impacts on poultry mite population as shown on Fig. 1. The accumulation of dust increases the ways for the mites to hide and anaemic unhealthy birds are more susceptible then to further attacks. Temperature and humidity also could play an important role (Nordenfors et al. 1999).

The current study in Italy observed that poultry breeds do not seem to influence the prevalence of this pest. Farms using the Warren breed and the Hy-line hybrid birds had a prevalence of infestation of 76.3 and 70.0%, respectively. Out of the 58 farms in the Italian study 65.5 and 34.5% were using the Warren and Hy-line breeds, respectively.

Discussion and conclusions

The results presented in Table 1 are comparable to those already published in other countries such as Kenya (Mungube et al. 2008) with 60% of *D. gallinae* infection in backyard chickens, Romania (Magdas et al. 2006) with a prevalence ranging from 57.5 to 72.5% depending of the locality, 100% in Poland (Cencek 2003), and 67% in Sweden (Höglund et al. 1995).

The diversity in terms of control methods and product used in some countries have shown the impact of the resistance capacity of the *Dermanyssus* populations (Marangi et al. 2008a) suggesting that an integrated method using more than one control methods could become the norm in many countries with acaricide restriction/resistance (Fiddes et al. 2005) to avoid recontamination of farm infrastructures knowing that mites can survive for a long time (Pavlićević et al. 2007) Such variation in acaricide resistance between countries could also explain the phylogenetic diversity between *D. gallinae* populations (Marangi et al. 2008b).

Temperature and season would also have an impact on poultry mite reports from farmers (Nordenfors et al. 1999). The fact that small farms have a higher level of infestation in Italy could be explained by the fact that on small premises farmers tend not to use air conditioning and do not have a break between production cycles allowing mites to feed on birds almost constantly.

Considering that this poultry pest can also attack other avian species (De Lope and Moller 1993; Gicik 1999; Romaniuk and Owczarzak-Podziemska 2002) could also boost prevalence rates in open poultry systems in which wild birds can enter and carry red mites. Even dogs, gerbils, rabbits, and other rodents have been observed carrying the poultry red mite allowing further import on farm infrastructures (Soulsby 1982; Bakr et al. 1995; Lucky et al. 2001).

It is also observed by colleagues in Serbia that another way to contaminate farms with the PRM is when farmers purchase used equipment. This situation will increase with the EU ban on conventional cages which could force many farmers who financially cannot buy new equipment to try reducing their costs by using infested second-hand enriched cages (personal communication, Pavlićević, Serbia).

Furthermore, an increase in the prevalence rates could also have an epidemiological impact on human and veterinarian diseases as the risks of *D. gallinae* transmitting more pathogens would increase as well (Valiente Moro et al. 2005, 2007, 2008).

The role of the national Veterinary Services (and veterinarians) is of paramount importance to assist farmers using the correct control products and dosing to avoid building even further acaricide resistance (see article from Mul and Koenraadt 2008). Knowledge transfer between veterinarians, scientists and the farming communities would also avoid misusing control methods which on a long term will bring more problems to the poultry industry.

It is also important to mention that due to the new EU Directive banning cages in 2012 some farmers are buying used equipment to reduce the costs or adapting to the new European legislation; doing so it increase the exchange of infested equipment passed between farms and contaminating new premises. It is therefore crucial for the poultry industry and the help of governments to constantly monitor mite population to put in place surveillance zones and movement restrictions when outbreaks of *D. gallinae* are observed on farms. In some European countries, such as in Italy, it is not compulsory to notify the Ministry of Agriculture when red poultry mite proliferations are observed on farms and it can lead to different attitudes from veterinarians. For this we would suggest an integrated and concerted European approach to report such infestations, which can spread between farms if good hygiene practices are not observed by the farm workers .

This paper has shown the importance and urgency linked to *D. gallinae* infestations and it is also important for governing bodies to participate in the control/eradication of such pest by funding networking and research collaborative work between industrials, researchers, and farmers.

Acknowledgments Dr. Olivier Sparagano would like to thank the BBSRC, the British Council and the Yorkshire Agricultural Society for partially sponsoring the work at Newcastle University. The work presented by colleagues from Italy and UK was also partially supported by a research grant (Ricerca Corrente 2006—Istituto Zooprofilattico Sperimentale della Puglia e della Basilicata) from the Ministry of Health, Italy and they would like to thank Professor Annunziata Giangaspero for her scientific inputs. The Dutch team would like to thank the Dutch Product Board of Poultry and Eggs for financing their study.

References

Anon (2003) Red mite confirmed in 60% of units. Poultry World, July

Auger P, Nantel J, Meunier N, Harrision RJ, Loiselle R, Gyorkos TW (1979) Skin acariasis caused by *Dermanyssus gallinae* (De Geer): an in-hospital outbreak. Can Med Assoc J 120:700–703

Bakr ME, Morsy TA, Nassef NEA, El-Meligi MA (1995) Mites infesting commensal rodents in Shebin El Kom, Menoufia G., Egypt. Egypt Soc Parasitol 25:853–859

Bellanger AP, Bories C, Foulet F, Bretagne S, Botterel F (2008) Nosocomial dermatitis caused by *Dermanyssus gallinae*. Infect Control Hosp Epidemiol 29:282–283. doi:10.1086/528815

Beugnet F, Chauve C, Gauthey M, Beert L (1997) Resistance of the red poultry mite to pyrethroids in France. Vet Rec 140:577–579

Cafiero MA, Circella E, Santagada G, Parisi A, Lomuto M, Camarda A (2007) Infestazione da *Dermanyssus gallinae* nell'uomo: un problema di igiene urbana. Obiet Doc Vet 6:41–45

Cafiero MA, Camarda A, Circella E, Santagada G, Schino G, Lomuto M (2008) Pseudoscabies caused by *Dermanyssus gallinae* in Italian city dwellers: a new setting for an old dermatitis. J Eur Acad Dermatol Venereol 22:1382–1383. doi:10.1111/j.1468-3083.2008.02645.x

Cencek T (2003) Prevalence of *Dermanyssus gallinae* in poultry farms in Silesia Region in Poland. Bull Vet Inst Pulawy 47:465–469

Chauve C (1998) The poultry red mite *Dermanyssus gallinae* (De Geer, 1778): current situation and future prospects for control. Vet Parasitol 79:239–245. doi:10.1016/S0304-4017(98)00167-8

Chirico J, Tauson R (2002) Traps containing acaricides for the control of *Dermanyssus gallinae*. Vet Parasitol 110:109–116. doi:10.1016/S0304-4017(02)00310-2

DeClercq J, Nachtegaele L (1993) *Dermanyssus gallinae* in a dog. Canine Pract 18:34–36

De Lope F, Moller A (1993) Effects of ectoparasites on reproduction of their swallow hosts—a cost of being multi-brooded. Oikos 67:557–562. doi:10.2307/3545368

Emous RA, van Fiks-van Niekerk TGCM, Mul MF (2005) €11 million damage for the sector: enquiry into the cost of mites to the poultry industry. De pluimveehouderij 35:8–9

Fiddes MD, Le Gresley S, Parsons DG, Epe C, Coles GC, Stafford KA (2005) Prevalence of the poultry red mite (*Dermanyssus gallinae*) in England. Vet Rec 157:233–235

Gicik Y (1999) Ectoparasites in wild pigeons on Ankara and vicinity. Kafkas Univ Vet Fak Derg 5:71–74

Guy JH, Khajavi M, Hlalele MM, Sparagano O (2004) Red mite (*Dermanyssus gallinae*) prevalence in laying units in Northern England. Br Poult Sci 45(2):S15–S16. doi:10.1080/00071660410001698001

Höglund J, Nordenfors H, Uggla A (1995) Prevalence of the poultry red mite, *Dermanyssus gallinae*, in different types of production systems for egg layers in Sweden. Poult Sci 74:1793–1798

Kristofik J, Masan P, Sustek Z (1996) Ectoparasites of bee-eater (*Merops apiaster*) and arthropods in its nests. Biologia 51:557–570

Lubac S, Dernburg A, Bon G, Chauve C, Zenner L (2003) Problématique et pratiques d'élevages en poules pondeuses dans le sud est de la France contre les nuisibles. poux rouges et mouches. In: ITAVI, INRA, AFSSA (eds) 5emes journées de la recherche avicole, Tours, France, 26–27 mars 2003, pp 101–104

Lucky AW, Sayers CP, Argus JD, Lucky A (2001) Avian mite bites acquired from a new source-pet gerbils. Arch Dermatol 137:167–170

Magdas C, Chirila F, Fit N, Criste A, Baciu H (2006) Epidemiologic study of *Dermanyssus gallinae* (Acari: Dermanyssidae) infestation in birds, from three localities on Cluj area. Bull Univ Agric Sci Vet Med 63:309–314

Marangi M, Cafiero MA, Capelli G, Camarda A, Sparagano OAE, Giangaspero A (2008a) Evaluation of the poultry red mite (*Dermanyssus gallinae*, Acarina: Dermanyssidae) susceptibility to some acaricides in a field population from Italy. Exp Appl Acarol (submitted to the same EAA special issue)

Marangi M, De Luna C, Cafiero MA, Camarda A, Le Bouquin S, Huonnic D, Giangaspero A, Sparagano OAE (2008b) Phylogenetic relationship between *Dermanyssus gallinae* populations in European countries based on mitochondrial CO1 gene sequences. Exp Appl Acarol (submitted to the same EAA special issue)

Mazyad SA, Abel El-Kadi M (2005) Ornithonyssus (Acari: Macronyssidae) mite dermatitis in poultry field-workers in Almarg, Qalyobiya governorate. J Egypt Soc Parasitol 35:213–222

Mignon B, Losson B (2008) Dermatitis in a horse associated with the poultry mite (*Dermanyssus gallinae*). Vet Dermatol 19:38–43

Mul MF, Koenraadt CJM (2008) Preventing introduction and spread of *Dermanyssus gallinae* in poultry facilities using the HACCP method. Exp Appl Acarol (submitted to the same EAA special issue)

Mungube EO, Bauni SM, Tenhagen BA, Wamae LW, Nzioka SM, Muhammed L, Nginyi JM (2008) Prevalence of parasites of the local scavenging chickens in a selected semi-arid zone of Eastern Kenya. Trop Anim Health Prod 40(2):101–109. doi:10.1007/s11250-007-9068-3

Nordenfors H, Höglund J, Uggla A (1999) Effects of temperature and humidity on oviposition, moulting and longevity of *Dermanyssus gallinae* (Acari: Dermanyssidae). J Med Entomol 36:68–72

Paoletti B, Iorio R, Traversa D, Gatti A, Capelli G, Giangaspero A, Sparagano OAE (2006) *Dermanyssus gallinae* in rural poultry farms in central Italy. In: XXVI national congress of the SOIPA, Messina, Italy, 21–24 June 2006. Parassitologia, vol 48, p 161

Pavlićević A, Pavlović I, Dotlić M (2007) A contribution to information on starvation survival capacity of poultry red mite *Dermanyssus gallinae*. Med Vet 50:485–491

Ramsay GW, Mason PC, Hunter AC (1975) Chicken mite (*Dermanyssus gallinae*) infesting a dog. N Z Vet J 23:155–156

Regan AM, Metersky ML, Craven DE (1987) Nosocomial dermatitis and pruritus caused by pigeon mite infestation. Arch Intern Med 147:2185–2187. doi:10.1001/archinte.147.12.2185

Romaniuk K, Owczarzak-Podziemska I (2002) The existence of *Dermanyssus gallinae* and saprobiontic mites in the litter of turkey farms. Med Weter 58:298–302

Rosen S, Yeruham I, Braverman Y (2002) Dermatitis in humans associated with the mites *Pyemotes tritici*, *Dermanyssus gallinae*, *Ornithonyssus bacoti* and *Androlaelaps casalis* in Israel. Med Vet Entomol 16:442–444. doi:10.1046/j.1365-2915.2002.00386.x

Sahibi H, Sparagano O, Rhalem A (2008) *Dermanyssus gallinae*: Acari parasite highly aggressive but still ignored in Morocco. In: BSP spring, trypanosomiasis/leishmaniasis and malaria meetings. March 30th, April 2nd, Newcastle Upon Tyne, p 173

Sexton DJ, Barton H (1975) Bird-mite infestation in a university hospital. Lancet 7904:445. doi:10.1016/S0140-6736(75)91506-8

Soulsby EJL (1982) Helminths, arthropods and protozoa of domesticated animals. Williams and Wilkins Co., Baltimore, pp 24–26

Valiente Moro C, Chauve C, Zenner L (2005) Vectorial role of some Dermanyssoid mites (Acari, Mesostigmata, Dermanyssoidea). Parasite 12:99–109

Valiente Moro C, Fravalo P, Amelot M, Chauve C, Zenner L, Salvat G (2007) Colonization and organ invasion in chicks experimentally infected with *Dermanyssus gallinae* contaminated by *Salmonella* Enteritidis. Avian Pathol 36:307–311. doi:10.1080/03079450701460484

Valiente Moro C, De Luna C, Guy JH, Sparagano OAE, Zenner L (2008) The poultry red mite, *Dermanyssus gallinae*, a potential vector of pathogenic agents. Exp Appl Acarol (submitted to the same EAA special issue)

Evaluation of the poultry red mite, *Dermanyssus gallinae* (Acari: Dermanyssidae) susceptibility to some acaricides in field populations from Italy

M. Marangi · M. A. Cafiero · G. Capelli · A. Camarda · O. A. E. Sparagano · A. Giangaspero

Originally published in the journal Experimental and Applied Acarology, Volume 48, Nos 1–2, 11–18.
DOI: 10.1007/s10493-008-9224-0

Abstract Red mite field populations from seven naturally infested Italian caged laying poultry farms were investigated for their susceptibility to acaricide formulations available on the market, containing amitraz, carbaryl and permethrin. A minimum of 3,000 mites of all stages were collected from each farm and were tested with five acaricide concentrations (5, 10, 20, 50, 100%) plus an untreated control (0%). Field red mite populations were found to be tolerant even with the highest concentrations with carbaryl and permethrin for six (86%) and three (42%) of the investigated farms, respectively ($P < 0.05$). Furthermore, six (86%) of the investigated farms showed a red mite population susceptible to amitraz at any concentration. Out of the seven field populations tested with amitraz, one population is becoming less tolerant whereas another was the most tolerant to carbaryl and permethrin at any concentration. Data show that the lack of effectiveness of some acaricides is spreading in Europe and call for the adoption of alternative management strategies to avoid development of resistance.

Keywords *Dermanyssus gallinae* · Carbaryl · Permethrin · Amitraz · Susceptibility · Italy

---

M. Marangi · A. Giangaspero (✉)
Dipartimento PrIME and Centro Interdipartimentale Bioagromed, Facoltà di Agraria, Università degli Studi di Foggia, 71100 Foggia, Italy
e-mail: a.giangaspero@unifg.it

M. A. Cafiero
Istituto Zooprofilattico Sperimentale della Puglia e Basilicata, Foggia, Italy

G. Capelli
Istituto Zooprofilattico Sperimentale delle Venezie, Padova, Italy

A. Camarda
Dipartimento di Sanità pubblica e Zootecnia, Università degli Studi di Bari, Bari, Italy

O. A. E. Sparagano
School of Agriculture, Food and Rural Development, Newcastle University, Newcastle upon Tyne, UK

Introduction

Arthropods are one of the most concerning threats in poultry industries due to both direct and indirect pathogenic effects. One of the most important arthropods affecting poultry production is in the genus *Dermanyssus* (Acari: Dermanyssidae) with *Dermanyssus gallinae* (De Geer) the most important (Fletcher and Axtell 1991; Maurer and Baumgärtner 1992).

Dermanyssus gallinae, also known as 'poultry red mite' or 'red mite' is a direct pest because it is an obligatory blood-sucking parasite; it is on the bird only to feed for 30–60 min and to spend the rest of the time concealed in cracks and crevices or in the litter. Its eggs are laid in the hiding places and hatch into six-legged larvae in 2–3 days. The chicken mite is seldom seen on the birds because of its intermittent feeding at night but skin lesions (especially on the breast and lower legs) are evidence of the feeding (Baker 1999).

In Europe, *D. gallinae* is one of the major problems in laying caged hens and less in broiler industries (Chauve 1998) due to the longer productive cycle related to the former management. The prevalence of infested farms ranges from 60% in UK (Guy et al. 2004; Fiddes et al. 2005) and Sweden (Hoglund et al. 1995) to 100% in Poland (Cencek 2003). A more recent prevalence study will be published in this special issue from several international teams showing the endemicity in several countries (i.e., UK, Italy, Serbia, Morocco, Japan, Montenegro and The Netherlands) of such poultry pest (Sparagano et al. 2008). It has been estimated that *D. gallinae* is costing the poultry industry in Europe a staggering 130 million euros per year (Van Emous 2005). In Italy, the huge numbers of parasites on the birds' cages and on the conveyor belts for eggs is well known; however, a prevalence data is only available in Abruzzo and Apulia regions with 20 and 74% of infestation in traditional (Paoletti et al. 2006) and industrial farms, respectively (Cafiero et al. 2008a).

Dermanyssus gallinae also occasionally bites mammals including man, and thus can constitute a problem of itching dermatitis to personnel working in affected poultry premises. Human cases have been also registered in Italy (Pampiglione et al. 2001) and several Red Mite Dermatitis (RMD) cases have been recently documented in city-dwellers in some Italian regions (Cafiero et al. 2007, 2008b) as well as in other countries (Green et al. 2007; Bellanger et al. 2008; Sahibi et al. 2008) and also on other animals (Ramsay et al. 1975; DeClercq and Nachtegaele 1993; Mignon and Losson 2008). *Dermanyssus gallinae* causes irritation, anaemia, and in some cases even death, and may result in blood-stained eggs and weight losses. Furthermore, *Dermanyssus* has been implicated in the transmission of several poultry pathogens such as viruses (chicken pox virus, Newcastle virus, fowl typhoid, and the agent of fowl cholera, Saint-Louis encephalitis virus), bacteria (*Francisella*, *Yersinia*, *Listeria*, *Pasteurella*, *Rickettsia*, *Salmonella*) and also parasites such as *Hepatozoon* (Zeman et al. 1982; Valiente Moro et al. 2005, 2008).

Once established in a flock, the control of red mites primarily relies on acaricide applications and relatively few are approved for the purpose worldwide. Carbamate, organophosphates (OPs), amidine and pyrethroid-based acaricides are the most widely used. Furthermore, many acaricides are not specifically labelled for use against red mites and if not properly applied the development of resistance can be accelerated. Due to the extensive and repeated usage of acaricides to control red mites, poultry farmers consider that the effectiveness of these molecules has diminished, possibly through the development of resistance (Thind and Ford 2007).

The less effectiveness of the acaricides has been also registered in Italy among farmers (Cafiero et al. 2008a, b) and an investigation on the reason of treatment failures has to be considered an urgent need.

In this study, we investigated red mite field populations from seven naturally infested Italian caged laying poultry farms—which had experienced problems in controlling mite populations—for their susceptibility to some molecules available on the market and more commonly used by farmers in the past and at present on livestock.

Materials and methods

From October 2007 to May 2008 seven caged laying hen poultry farms were investigated to evaluate the efficacy of amitraz, carbaryl and permethrin. Five farms (A–E) were in the South and two (F, G) in the North of Italy. A minimum of 3,000 mites of all stages were collected near the feeder and along the conveyor belt from each farm. In the laboratory the mites were starved in the dark at room temperature for 1 week to allow the full digestions of blood meal.

Three acaricides Amitraz 100 ml/l (TakTik125, Gellini®), Carbaryl 47.5 g kg^{-1} (AS50, Sipcam®) and Permethrin 0.5 g kg^{-1} (Pertrin polvere, Copyr®) available on the Italian market were used and processed following the procedures described by Thind and Muggleton (1998). For each acaricide aliquots of the stock solution were diluted to provide a range of five concentrations (dose points) which were used to impregnate Whatman n. 1 filter papers. Using a 1 ml pipette, an aliquot of 0.2 ml of the required pesticide concentration, was spread on each filter paper using a progressively decreasing spiral motion to ensure an even distribution of the pesticide on the filter paper. After treatment the filter papers were removed by pins to prevent cross-contamination or loss of pesticide. The treatment of the filter papers was carried out in ascending order of pesticide concentration. The concentrations used were expressed as micrograms of active ingredient per square centimeter (μg cm^{-2}). For the three acaricides the six concentrations were 100, 50, 20, 10, 5, and 0%.

Twenty mites of all stages were placed in the depression of the filter paper of the test cell (Thind and Muggleton 1998). The mites were confined in the test cell for 24 h, after which the test cells were opened and live and dead mites were counted. The mites were considered alive if they exhibited any form of movement, with or without stimulation from the single hairbrush. One hundred and twenty mites, in batches of 20, were exposed to each concentration of acaricides and to control papers treated with only water. The experiment was made over three days in the same lab conditions (Humidity: 60%; Temperature: 20°C).

The efficacy (%) of each test was evaluated using the following formula: percent age = (number of live mites in control − number of live mites in treated)/number of live mites in control × 100. For each formulation, the mean efficacy was compared according to site and concentration of active compound using one-way ANOVA, followed by the Tukey test for paired post-hoc comparison.

Results

Amitraz showed an efficacy of 100% for each concentration in all farms investigated with the exception of Farm G in which the efficacy was significantly lower for the 5 and 10% concentrations ($P < 0.05$) (Table 1).

Carbaryl did not achieve a good mean efficacy at any concentration in Farms A, B, C, D, F and G, while it reaches a good efficacy (95%) only in Farm E although at the highest concentration (100%) (Table 2; Fig. 1).

Table 1 Comparison of the mean efficacy ($n = 6$) of five concentrations of Amitraz against the poultry red mite from seven farms in Italy

Farms	5%	10%	20%	50%	100%
A	100	100	100	100	100
B	100	100	100	100	100
C	100	100	100	100	100
D	100	100	100	100	100
E	100	100	100	100	100
F	100	100	100	100	100
G	79.63	83.33	94.44	100	100
ANOVA *P*	0.017	0.018	0.511		

The efficacy of permethrin was good (from 95 to 100%) in Farm C but only at the highest concentrations (50 and 100%, respectively) while the efficacy reaches 95 and 93% at the concentration of 100% in Farms D and E, respectively (Table 3; Fig. 2).

The differences among the farms were statistically significant only starting at the highest concentration (up to 20%) of both carbaryl and permethrin and the diversity is shown in Tables 2, 3.

Discussion

This study shows that, in Italy, field red mite populations are tolerant even at the highest concentrations to carbaryl and to permethrin for 6 (86%) and 3 (42%) of the investigated farms, respectively. Furthermore, 6 (86%) of the investigated farms showed a red mite population susceptible to amitraz at any concentration.

Out of the seven tested field populations with amitraz, the G population is the only one becoming susceptible while the population A was the most susceptible to carbaryl and permethrin at any concentration; the negative patterns showed by Farm A for both carbaryl and permethrin (Figs. 1, 2) is related to the high mortality of control mites due to unknown reason, being all the tests run in the same conditions and time; however, the data are not statistically significant to get conclusions.

Table 2 Comparison of the mean efficacy ($n = 6$) of five concentrations of carbaryl against the poultry red mite from seven farms in Italy

Farms	5%	10%	20%	50%	100%
A	−12.50	−9.37	-3.12^{cd}	0.00^{dgmn}	9.37^{ab}
B	21.17	19.47	8.36^{a}	19.21^{ach}	55.42
C	4.28	30.32	41.58^{d}	76.34^{efg}	88.09^{bd}
D	21.39	24.16	30.27	49.58^{n}	52.91
E	2.36	4.30	51.94^{abc}	82.50^{abcd}	95.55^{af}
F	−2.96	28.24	28.24	44.44^{bm}	42.77
G	9.60	9.70	7.39^{b}	17.57^{cf}	21.50^{df}
ANOVA *P*	0.137	0.289	0.002	0.000	0.003

An equal letter following means within a column indicates a statistically significant difference ($P < 0.05$)

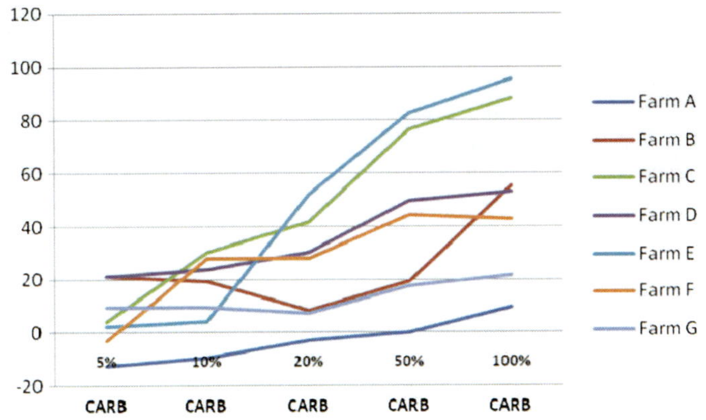

Fig. 1 Efficacy (%) of five concentrations of carbaryl against poultry red mite from seven poultry farms

Table 3 Comparison of the mean efficacy ($n = 6$) of five concentrations of permethrin against the poultry red mite from seven farms in Italy

Farms	5%	10%	20%	50%	100%
A	−15.62[a]	−3.12	−3.12[bfgh]	25.00[de]	28.12[bilm]
B	29.02	40.69	53.89[ab]	63.61	75.55[ab]
C	32.33	46.43	65.12[hl]	95.83[be]	100[gm]
D	50.00[a]	62.50	68.75[df]	66.66	95.83[ei]
E	19.55	25.68	50.99[g]	77.82[ad]	93.99[fl]
F	ND	ND	ND	ND	ND
G	17.73	15.98	23.19[adl]	26.61[ab]	30.31[aefg]
ANOVA P	0.056	0.098	0.000	0.002	0.000

ND Not determined. An equal letter following means within a column indicates a statistically significant difference ($P < 0.05$)

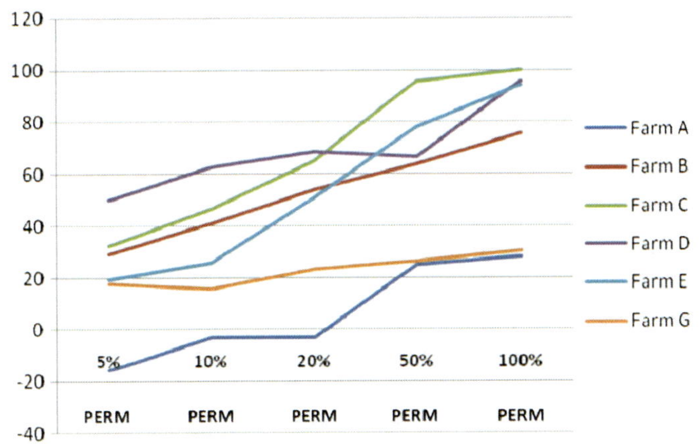

Fig. 2 Efficacy (%) of five concentrations of permethrin against poultry red mite from six poultry farms

In Italy, as well as in other European countries, besides to physical methods, current control methods rely on the use of acaricides, i.e., OPs, carbamates and pyrethroids.

Although the efficacy of several acaricides have been tested in vitro against *D. gallinae* using different techniques (Fletcher and Axtell 1991; Zeman and Zelezný 1985; Zeman 1987; Abo-Taka 1990; Beugnet et al. 1997), none of them are specifically labelled for use against red mites. Moreover, some are unsuitable from a food safety point of view and for environmental reasons and some are now banned by the EU (such as carbamates).

It is possible that, due to the lack of licensed molecules against *D. gallinae*, for the prophylaxy and control of this poultry pest, Italian poultry farmers have very little choice and very often use acaricides licensed for crop agriculture and/or livestock pest control instead. The results on the tolerance observed in this study, and in particularly for carbaryl, seems to confirm this hypothesis. Also, it is not excluded that farmers may often manage acaricides incorrectly (e.g., wrong dilution of active ingredients, wrong treatment schedule, low concentration) and, consequently be responsible of promoting a more quickly development of acaricides resistance in red mites.

Resistance to pyrethroids have been detected in Northern Europe such as UK (Fiddes et al. 2005), Sweden (Nordenfors et al. 2001) and in France (Beugnet et al. 1997) as well as resistence to carbamates and pyretroids have been observed in Germany (Liebish and Liebish 2001). Also, resistence to DDT, organophosphates or pyrethroids has been suspected in the former Czechoslovakia (Zeman and Zelezný 1985) and in Italy since in the 1980s (Genchi et al. 1984). This study is the first report of tolerance to carbamate and pyrethroid in a field population of *D. gallinae* from Northern to Southern Italy.

The results obtained in this study are troublesome; in fact, the less effectiveness of the acaricides registered in the investigated farms can confirm the extensive and improper usage by farmers of the acaricides. In this sense, considering that the dosage of amitraz (TakTik125, Gellini®) for livestock (swine) against the mange is of 0.05% p.a., the percentage of mite mortality of 79.63 and 83.33% registered in this study at 5 and 10% of amitraz concentration, respectively (Table 1), is of great concern.

These data show that the lack of effectiveness of some acaricides is spreading more and more in Europe. Consequently, these data reinforce the need for regular monitoring campaigns on the diffusion of red mites and potential resistance throughout Italy. Actually, it may be desirable that the National Health Authority and commercial companies might intervene in this field, soliciting and favouring the registration of effective and safe molecules as the improper use of unlicensed products would increase the possibility that residues may be accumulated in poultries' eggs, meat, liver and fat.

However, it must be underlined that, due to the lack of official cut-off for the evaluation of the efficacy of acaricides against poultry mites, a recognized guidelines would be needed.

In conclusion, the existence of possible resistant poultry red mite *D. gallinae* populations in Italy and in other countries is a very critical problem. This situation requires that both scientists and industries develop further research in this field and investigate new control methods. Furthermore, waiting for approved and safe molecules in the poultry industry could be counterproductive and Italian poultry farmers should be pressed by Health Institutions to adopt alternative management strategies (Chauve 1998) rather than relying completely on chemical control methods.

Acknowledgments This work was supported by a research grant (Ricerca Corrente 2006—Istituto Zooprofilattico Sperimentale della Puglia e Basilicata) from the Ministry of Health, Italy and by a grant (British-Italian Partnership Programme) from the British Council-CRUI, Italy.

References

Abo-Taka SM (1990) Toxicity of certain pesticides against the chicken mite *Dermanyssus gallinae*. Med Vet Entomol 4:125–126. doi:10.1111/j.1365-2915.1990.tb00268.x

Baker AS (1999) Mites and ticks of domestic animals. An identification guide and information source. The Stationery Office, The Natural and History Museum, London 240 pp

Bellanger AP, Bories C, Foulet F, Bretagne S, Botterel F (2008) Nosocomial dermatitis caused by *Dermanyssus gallinae*. Infect Control Hosp Epidemiol 29:282–283. doi:10.1086/528815

Beugnet F, Chauve C, Gauthey M, Beert L (1997) Resistance of the poultry red mite to pyrethroids in France. Vet Rec 140:577–579

Cafiero MA, Circella E, Santagada G, Parisi A, Lomuto M, Camarda A (2007) Infestazione da Dermanyssus gallinae nell'uomo. Un problema di igiene urbana. Obiettivi Documenti Veterinari vol 6, pp 41–45

Cafiero MA, Camarda A, Circella E, Raele D, Giangaspero A, Sparagano OAE (2008a) Prevalence in laying hen units in the Apulia region. Italy In: Proceeding of the British society for parasitology, 30th March–2nd April 2008, Newcastle, p 170

Cafiero MA, Camarda A, Circella E, Santagada G, Schino G, Lomuto M (2008b) Pseudoscabies caused by *Dermanyssus gallinae* in Italian city dwellers: a new setting for an old dermatitis. J Eur Acad Dermatol Venereol. doi:10.1111/j.1468-3083.2008.0264.x

Cencek T (2003) Prevalence of *Dermanyssus gallinae* in poultry farms in Silesia Region in Poland. Bull Vet Inst Pulawy 47:465–469

Chauve C (1998) The poultry red mite *Dermanyssus gallinae* (De Geer 1778): current situation and future prospects for control. Vet Parasitol 79:239–245. doi:10.1016/S0304-4017(98)00167-8

DeClercq J, Nachtegaele L (1993) *Dermanyssus gallinae* in a dog. Canine Pract 18:34–36

Fiddes MD, Le Gresley S, Parsons DG, Epe C, Coles GC, Stafford KA (2005) Prevalence of the poultry red mite (*Dermanyssus gallinae*) in England. Vet Rec 157:233–235

Fletcher MG, Axtell RC (1991) Susceptibility of northern fowl mite, *Ornithonyssus sylviarum* (Acarina: Macronyssidae), and chicken mite, *Dermanyssus gallinae* (Acarina: Dermanyssidae) to selected acaricides. Exp Appl Acarol 13:137–142. doi:10.1007/BF01193664

Genchi C, Huber H, Traldi G (1984) The efficacy of flumethrin (Bayticol Bayer) for the control of chicken mite *Dermanyssus gallinae* (De Geer 1778) (Acarina, Dermanyssidae). Arch Vet Ital 35:125–128

Green B, Sparling J, Sparling MB (2007) What's eating you? pigeon mite (*Dermanyssus gallinae*). Cutis 80:461–462

Guy JH, Khajavi M, Hlalele MM, Sparagano O (2004) Red mite (*Dermanyssus gallinae*) prevalence in laying units in Norther England. Br Poult Sci 45:15–16. doi:10.1080/00071660410001698001

Hoglund J, Nordenfors H, Uggla A (1995) Prevalence of the poultry red mite, *Dermanyssus gallinae*, in different types of production systems for egg layers in Sweden. Poult Sci 74:1793–1798

Liebish A, Liebish G (2001) Bekämpfung der Roten Vogelmilbe. Teil 1. Resistenzen und Prufung neuer Mittel bei der Bekämpfung von Roten Vogelmilbe (*Dermanyssus gallinae*). Schadlingsbekampfer 3:11–13

Maurer V, Baumgärtner J (1992) Temperature influence on life table statistics of the chicken mite *Dermanyssus gallinae* (Acari: Dermanyssidae). Exp Appl Acarol 15:27–40. doi:10.1007/BF01193965

Mignon B, Losson B (2008) Dermatitis in a horse associated with the poultry mite (*Dermanyssus gallinae*). Vet Dermatol 19:38–43

Nordenfors H, Höglund J, Tauson R, Chirico J (2001) Effect of permethrin impregnated plastic strips on *Dermanyssus gallinae* in loose-housing systems for laying hens. Vet Parasitol 102:121–131. doi:10.1016/S0304-4017(01)00528-3

Pampiglione S, Pampiglione G, Pagani M, Rivasi F (2001) Infestazione persistente del cuoio capelluto da *Dermanyssus gallinae* in una contadina emiliana. Parassitologia 43:113–115

Paoletti B, Iorio R, Traversa D, Gatti A, Capelli G, Sparagano OAE, Giangaspero A (2006) Dermanyssus gallinae in rural poultry farms in Central Italy. Atti XXIV Congresso Soc Ital Parassitologia, vol 48, p 161

Ramsay GW, Mason PC, Hunter AC (1975) Chicken mite (*Dermanyssus gallinae*) infesting a dog. N Z Vet J 23:155–156

Sahibi H, Sparagano O, Rhalem A (2008) *Dermanyssus gallinae*: Acari parasite highly aggressive but still ignored in Morocco. BSP spring, Trypanosomiasis/Leishmaniasis & Malaria meetings. March 30th, April 2nd, Newcastle upon Tyne, UK, p 173

Sparagano O, Pavlićević A, Murano T, Camarda A, Sahibi H, Kilpinen O, Mul M, Van Emous E, Le Bouquin S, Hoele S, Cafiero MA (2008) Prevalence and key figures for the poultry red mite *Dermanyssus gallinae* infections in poultry farm systems. Exp Appl Acarol (submitted to the same special issue)

Thind BB, Ford HL (2007) Assessment of susceptibility of the poultry red mite *Dermanyssus gallinae* (Acari: Dermanyssidae) to some acaricides using an adapted filter paper based bioassay. Vet Parasitol 144:344–348. doi:10.1016/j.vetpar.2006.10.002

Thind BB, Muggleton J (1998) A new bioassay method for the detection of resistance to pesticides in the stored product mite *Acarus siro*. Exp Appl Acarol 22:543–552. doi:10.1023/A:1006037807498

Valiente Moro C, Chauve C, Zenner L (2005) Vectorial role of some dermanyssoid mites (Acari, Mesostigmata, Dermanyssoidea). Parasite 12:99–109

Valiente Moro C, De Luna C, Guy JH, Sparagano OAE, Zenner L (2008) The poultry red mite, *D. gallinae*, a potential vector of pathogenic agents. Exp Appl Acarol (submitted to the same special issue)

Van Emous R (2005) Wage war against the red mite! Poul Int 44:26–33

Zeman P (1987) Encounter the poultry red mite resistance to acaricides in Czechoslovak poultry-farming. Folia Parasitol (Praha) 34:369–373

Zeman P, Zelezný J (1985) The susceptibility of the poultry red mite, *Dermanyssus gallinae* (De Geer 1778), to some acaricides under laboratory conditions. Exp Appl Acarol 1:17–22. doi:10.1007/BF01262196

Zeman P, Stika V, Skalka B, Bártík M, Dusbábek F, Lávicková M (1982) Potential role of *Dermanyssus gallinae* De Geer, 1778 in the circulation of the agent of pullurosis-typhus in hens. Folia Parasitol (Praha) 29:371–374

Exploration of the susceptibility of AChE from the poultry red mite *Dermanyssus gallinae* (Acari: Mesostigmata) to organophosphates in field isolates from France

Lise Roy · Claude Chauve · Jean Delaporte · Gilbert Inizan · Thierry Buronfosse

Originally published in the journal Experimental and Applied Acarology, Volume 48, Nos 1–2, 19–30.
DOI: 10.1007/s10493-009-9249-z

Abstract The red fowl mite *Dermanyssus gallinae* (De Geer, 1778) is a hematophagous mite species, which is very commonly found in layer facilities in Europe. The economic and animal health impact of this parasite is quite important. In laying hen houses, organophosphates are almost the only legally usable chemicals. Detecting a target resistance can be useful in order to limit the emergence of resistant populations. The acetylcholinesterase (AChE) activity and the enzyme sensitivity to paraoxon was investigated in 39 field samples and compared to a susceptible reference strain (SSK). Insensitivity factor values (expressed as IC_{50} ratio) obtained from field isolates compared to SSK revealed some polymorphism but not exceeding a 6-fold difference. The kinetic characteristics of AChE from some field samples showed some difference in K_M values for acetylthiocholine and inhibition kinetics performed with diethyl paraoxon exhibited a 5.5-fold difference in the bimolecular rate constant in one field isolate. Taken together, these data suggested that differences in AChE susceptibility to organophosphates may exist in *D. gallinae* but no resistant population was found.

Keywords *Dermanyssus gallinae* · Inhibition of acetylcholinesterase · Paraoxon · Field isolates · Organophosphate resistance

Introduction

Dermanyssus gallinae (De Geer, 1778) or the poultry red mite is a cosmopolitan hematophagous mite, parasitic on birds. Five life stages are known for this species (egg, larva, protonymph, deutonymph, adult), two of which need a blood meal for performing

L. Roy (✉) · C. Chauve · T. Buronfosse
Ecole Nationale Vétérinaire de Lyon, Laboratoire de parasitologie, Université de Lyon, 69280 Marcy-L'Etoile, France
e-mail: l.roy@vet-lyon.fr

J. Delaporte · G. Inizan
Bayer Healthcare Animal Health, Bayer Pharma S.A.S, 10A, rue Joseph Le Brix, BP 40011, 56891 Saint-Ave Cedex, France

metamorphosis (protonymph, deutonymph; Wood 1917). Adult females need blood meals for egg maturation. The economic impact of this parasite is quite important with more or less serious direct damages, such as anemia (Kirkwood 1967; Keçeci et al. 2004), possible death from exsanguination, decreased egg production, but also possible transmission of certain bacterial or viral diseases (avian spirocheatosis, fowl cholera, salmonellosis, etc.) (Valiente Moro et al. 2005, 2007). Moreover, some well visible blood spots on egg shells induce a heavy financial loss with downgraded eggs. It is especially injurious in layer houses in Europe and, today, controlling the spread of these mites is an economic challenge. Because of the Maximum Residue Limits in eggs, only few products are allowed for the control of *D. gallinae* in Europe.

Insects and other arthropods have developed different mechanisms to escape to a selective pressure imposed by the use of the same insecticide. One of the adaptive mechanisms, which confer resistance to organophosphates (OPs) and carbamate pesticides, is allowed by a modification of the acetylcholinesterase (AChE), the insecticide target protein (Fournier and Mutero 1994). AChE catalyses the hydrolysis of the neurotransmitter, acetylcholine, thereby ending transmission of nerve impulses at the synapses of cholinergic neurones. The inhibition of this enzyme leads to paralysis and death of arthropods. Conversely, AChEs that are not inhibited by OPs and carbamates confer resistance to these pesticides. In resistant arthropods, a structural change of AChE preserves it from being inhibited by the OPs. This structure modification has been shown in more than 33 insect and acari species (Fournier and Mutero 1994). For instance: some crop pests such as spider mites (Stumpf et al. 2001; Tsagkarakou et al. 2002) or codling moth (Reuveny and Cohen 2004), some stored-food pests such as Psocidae (Wang et al. 2004), some potential diseases vectors such as mosquitoes (Weill et al. 2003), some ticks (Stone et al. 1976; Baxter et al. 1999; Pruett 2002), etc. The molecular basis of this resistance has been characterized for some insects (Zhu et al. 1996; Newcomb et al. 1997; Nabeshima et al. 2004) and ticks (Xu et al. 2003) and is associated with specific mutations in the *ace* genes. Single or multiple amino acid substitutions confer distinct catalytic properties to the mutated protein leading to a decreased sensitivity of AChE to inhibition by OPs insecticides.

This phenomenon has been developed following extensive and prolonged use of these insecticide compounds. Because the frequency of *D. gallinae* infestations is currently increasing and that large populations can be established rapidly under favorable conditions, farmers, worried by economic losses, use chemical acaricide treatments at least in the empty chicken houses (Chauve 1998). All these factors are therefore prerequisite to expect that a resistant strain may be favored if a benefit mutation point in the *D. gallinae* AChE gene arised. High levels of resistance in *D. gallinae* have been reported for DDT and permethrin resistant mites was suspected to be involved as the main reason for the failure to control some *D. gallinae* populations (Zeman and Zelezny 1985; Nordenfors et al. 2001).

In French laying hen houses, almost only OP compounds can be legally used, and, until recently, only between flocks. Previously, no ectoparasiticide was allowed to be used during flocks except some products composed of vegetal extracts, inert substances and some detergents, with mechanical actions. The lack of efficacy of these compounds compared to cholinesterase inhibitors incitated egg farmers to use these chemicals as soon as the poultry red mites represented a major problem in aviary systems. As a result, many populations of European mites might have been repeatedly exposed to OPs. The fast development potential of *D. gallinae* in layer houses conditions and the applications of OPs, legally done between flocks and sometimes illegally during flocks, to maintain the mite populations below economic thresholds are factors that may have facilitate the emergence of insecticide resistance in this species. Since 2007, an OP ectoparasiticide, phoxim, which can assure a 0-day withholding period for eggs, was approved by EMEA to be used, in Europe, to treat a *D. gallinae* infestation in poultry houses stocked with egg-laying hens (Keïta et al. 2006). Thus, one can suppose that as a new OP-based product is going to be used during flocks, poultry red mites will be exposed constantly, forward favoring the emergence of resistant populations. The aim of the present study, which was conducted before the commercial authorization for using phoxim, was to use a biochemical assays for monitoring AChE in mites coming from different layer houses from different French counties in order to investigate the possible existence of resistant strains.

Materials and methods

Mites

Fourty different populations of mites were used in this study. A putative susceptible reference strain called Standard Strain Kilpinen (SSK) was kindly provided by Dr Ole Kilpinen (Lyngby, Denmark). This strain has been cultured in laboratory conditions since 1997 and has not been exposed to OPs since at least that date.

The 39 other mite populations were collected in 39 independent layer houses from various counties in France. All isolates were maintained alive separately in the laboratory for few days (less than 6 days) allowing the emergence of a sufficient number of protonymphs to perform the biochemical assay. Thus, as mites were directly coming from farms, most of protonymphs were engorged. In order to get enough living and unengorged protonymphs, females were allowed laying their eggs and eggs hatching by placing each strain at room temperature in an open box. Each box was placed into a large bowl filled with water (with a drop of a tension-active agent) so that mites cannot escape and that inter-isolate contamination is avoided.

To standardize the biochemical assay as a potential diagnostic test and to avoid expected interferences due to blood meal in adults or in deutonymphs, only unengorged protonymphs were ground and AChE extracted. Typically, 200 protonymphs were placed into a 1.5 ml microcentrifuge tube and killed by freezing (−20°C for 2 h). Samples were ground in ice in the same tube containing 1.4 ml of a 10 mM pH 7.5 Tris–HCl buffer, 5.84% (m/v) NaCl, 0.4% Triton X-100 and 25 mM EDTA using a Potter's device. AChE was extracted for 20 min at 4°C and extracted AChE was harvested in the supernatant following centrifugation at 14.000*g* for 10 min at 4°C. These extracts were immediately used for measuring AChE activity.

AChE assays

Basic principle

AChE activity of protonymph extracts was measured with a modification of the Ellman assay based on the enzymatic hydrolysis of acetylthiocholine iodide, ASCh, (Ellman et al. 1961). Reactions were conducted in 96-well microplates (Maxisorp, Nunc, France). Typical AChE activity of protonymph extract was assayed on 100 μl of extract mixed with 100 μl of buffering solution containing Tris-HCl 0.5 mM pH 8.0, 1.6 mM of 5-5′-dithio-bis

(2-nitro-benzoic acid) (DTNB, Sigma chemicals) and ASCh 1 mM. The change of absorbance at 410 nm was measured every 4 min for 40 min at 25°C on a Dynex microplates reader. Each assay was performed in duplicate. The spontaneous ASCh hydrolysis was corrected to the signal obtained from each well by subtracting the change of absorbance occurring in wells where protonymph extracts were omitted.

Preliminary assays for the validation of the test

Because assays on the susceptibility of AChE have never been done on *D. gallinae*, some parameters were first checked for the validation of tests. The linearity of the reaction with time and with the enzyme amount has been explored by measuring the enzymatic activity on extracts containing various numbers of protonymphs (from 50 to 400 protonymphs) in the same amount of lysis buffer.

The stability of protonymph AChE to freezing was evaluated over a 3 months period using several aliquots of protonymphs conserved at −20°C. These aliquots were regularly extracted and their AChE activity measured between day 1 and 90.

Moreover, AChE activity was measured exclusively on this stage because other stages (deutonymphs or adults) may content some residual esterase activities coming from the host's blood. However, few assays were conducted on AChE extracted from adults after 2 weeks of starvation. Measured activities were roughly in the same order of magnitude than those observed with protonymph extracts but interassay variations appeared to be important (results not shown), suggesting a possible interaction of remaining blood enzyme activities as soon as mites have had a blood meal.

In order to verify that the change in the absorbance at 410 nm of protonymph extracts was dependent of AChE activity, inhibition studies were assayed. The thermal inactivation of the enzymatic activity was evaluated by measuring the residual activity of the extract after an incubation for 10 min at 50°C of the protonymph extracts. The ability of the carbamate eserine sulphate to inhibit the change in the absorbance was also evaluated by measuring the activity in the presence of 0.5 mM of eserine.

Exploration of kinetic parameters of extracted AChE

The kinetic parameters K_M and V_{max} of AChE extracts from the SSK strain and four isolates from field were determined with 12 different concentrations of ASCh ranging from 10 μM to 5 mM of final concentration. The AChE activity was converted to picomoles of ASCh hydrolyzed per min and per protonymph using 1.36 10^4 M^{-1} cm^{-1} as molar extinction coefficient. Kinetic constants K_M and V_{max} were obtained by linear regression after fitting a Lineweaver-Burk double-reciprocal plot of the Michaelis-Menten function.

OP inhibitory assay

To investigate whether the AChE activity, extracted from protonymphs sampled in different laying farms, was susceptible to OP inhibition, inhibition kinetics were analyzed with the OP inhibitor diethyl paraoxon which is the form of the insecticide that irreversibly inhibits AChE. The organophosphate diethyl paraoxon [*O*,*O*-diethyl-*O*-(4-nitrophenyl) phosphate] was purchased from Riedel-de Haen, Seelze, Germany and was used in inhibitory assays as an AChE specific inhibitor.

Progressive inhibition of AChE activity by OP over time

A progressive inhibition of AChE activity over time was performed in the AChE extract from the SSK strain and three additional field samples (number 6002, 6005 and 22S84). The inhibitory action of diethyl paraoxon on AChE extracts from these samples was also analyzed by calculating the bimolecular rate constant, k_i, using the Aldridge method (1950). Briefly, residual AChE activity was measured as stated above for standard AChE assay except that supernatant extract was incubated with different concentrations of diethyl paraoxon (2.0×10^{-7} M, 1.0×10^{-7} M, 5.0×10^{-8} M) for various times (0, 6, 12, 19 and 30 min) prior the addition of the substrate reagent. In each assay, blank controls were done with wells without protonymph extract, in order to subtract non enzymatic hydrolysis of ASCh. The logarithm of the residual activity was plotted against the preincubation time and the bimolecular rate constant, k_i, was extracted by linear regression by dividing the slope by the inhibitor concentration in accordance with Aldridge (1950).

Screening of 39 field populations

In order to explore potential variations in the susceptibility to OP inhibition of AChE extracted from protonymphs sampled in different laying farms, inhibition of AChE for each 39 field samples was measured in the presence of 1 mM ASCh by three concentrations of inhibitor and without any preincubation time. Paraoxon concentrations used were 8.0×10^{-7} M, 2.0×10^{-7} and 5.0×10^{-8}. These concentrations were selected to inhibit between 20 and 80% of the residual AChE activity in SSK strain without any delay between the addition of diethyl paraoxon and the addition of the substrate reagent. Each sample was analyzed in two separate assays. In each duplicate, a reaction without the inhibitor was included as a control. An OP inhibitory assay in duplicate of the SSK strain was also systematically included in each plate. All the assay conditions were the same as that used for kinetics assays. The inhibitory data were analyzed by plotting residual activities with the inhibitor concentration. Inhibitory concentration 50% (IC_{50}) value of each field sample was evaluated by linear regression and the ratio of this IC_{50} value with that obtained with the SSK strain in the same plate was calculated.

Statistical analysis

Kinetic parameters were compared using a non-parametric Kruskal–Wallis test.

Results

Validation of AChE extraction

The AChE activity was measured from an extract of protonymphs from the SSK strain in the supernatant and in the pellet after suspending the disrupted protonymph fragments in 100 μl of the lysis buffer. Variations in the change of absorbance obtained when incubations were performed with the pellet were not significantly higher than the spontaneous hydrolysis of ASCh, suggesting that no AChE activity remained present in the non-extracted fraction (results not shown). The AChE activity was linear with incubation time up to 40 min and with the amount of protonymphs extracted up to 200 protonymphs per 1.4 ml of lysis buffer (Fig. 1a). Above this quantity, the activity plateaued probably due to

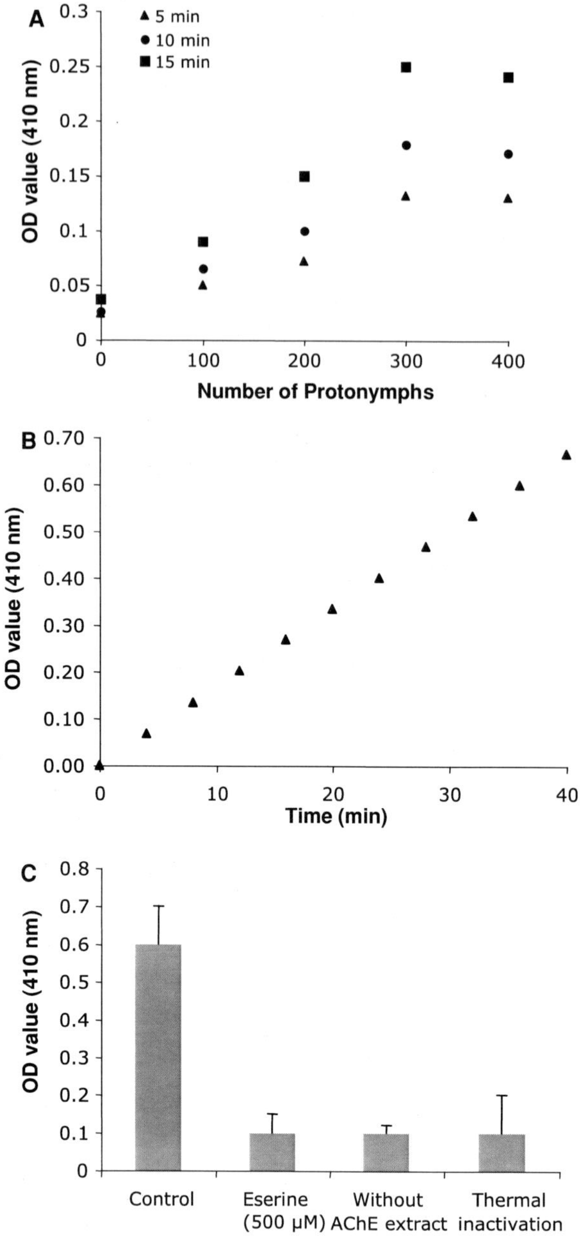

Fig. 1 Kinetic properties of acetylcholinesterase extracted from N1 stage from *Dermanyssus gallinae*. **a** Acetylcholinesterase activity of *D. gallinae* extracted from different quantities of protonymphs. Each data point represents the velocity of thiocholine production from acetylthiocholine calculated as the OD (410 nm) changes after 5, 10 or 15 min incubation time in the presence of the substrate (1 mM). **b** Hydrolysis of acetylthiocholine as a fonction of incubation time when AChE extracted from 200 protonymphs is used. **c** Acetylcholinesterase activity in the presence or absence of eserine (0.5 mM) or when extracts containing acetylcholinesterase have been incubated 10 min at 50°C prior the introduction of the substrate. OD (410 nm) was measured after 40 min of incubation. *Bars* represent mean ± SD

competitive reactions. For this reason, all assays were further performed using 200 protonymphs per extraction. The OD curve was linear with incubation time at least up to 40 min (Fig. 1b).

The catalytic activity mediated by the protonymph extract was totally inhibited by thermal pretreatment at 50°C for 10 min or by incubation with 0.5 mM eserine sulphate, a carbamate compound that specifically inhibits AChE (Fig. 1c).

Further, AChE stability to freezing was checked over a 90-days period. No significant loss of AChE activity was observed in protonymphs that had been stored at −20°C for up to 3 months before extraction (data not shown). Thus, this microtiter plate assay using *D. gallinae* protonymphs was proved to be suitable for measuring AChE activity and performing inhibition studies.

Kinetic parameters of AChE in *Dermanyssus gallinae*

The catalytic properties of *D. gallinae* AChE extracts have been characterized in five different strains coming from a laboratory source (SSK) or from field samples. Kinetics of the AChE extracts using ASCh as artificial substrate followed Michaelis–Menten kinetics. Kinetic parameters of the AChE extracts are shown in Table 1. Despite apparent differences in K_M values between isolates under test, there was no significant difference. Nevertheless, a lower V_{max} value was obtained with AChE extracts from SSK and 6001 ($P = 0.004$) compared to the other field samples.

The AChE inhibition kinetics obtained with the reference strain SSK and three field samples (numbers 6002-6005-22S84) were further characterized for their sensitivity to diethyl paraoxon. The progressive inhibition of AChE curves followed a pseudo first order kinetics (Fig. 3). The apparent bimolecular rate constants (k_i) for AChE inhibition were extracted for each of these samples and values are given in Table 2. The k_i values for the 6005 and 22S84 field samples were in the same range of magnitude than that obtained with AChE extracted from SSK strain whereas the bimolecular rate constant for 6002 strain was significantly different ($P < 0.01$) from the sensitive reference strain (Table 2). The lower k_i

Table 1 K_M and V_{max} values of AChE extracted from *Dermanyssus gallinae* protonymphs coming from SSK strain and from different field samples

Strain/field isolates	V_{max} (pmol^{-1} min^{-1} protonymph)	K_M (μM)
SSK ($n = 8$)	62.1 ± 8.7	36.4 ± 9.5
22S84 ($n = 6$)	106.5 ± 9.6	54.8 ± 18.5
35S72 ($n = 3$)	87.0 ± 4.1	44.7 ± 12.0
6001 ($n = 3$)	57.2 ± 1.1	38.7 ± 9.0
Berthet ($n = 3$)	105.0 ± 9.9	52.7 ± 7.6

Values are expressed as mean ± SD, numbers in parentheses represent number of independent determinations

Table 2 Bimolecular rate constants (k_i) for AChE extracts inhibition by paraoxon in SSK strain and 3 field samples

Strain/field isolates	Bimolecular rate constant (M^{-1} min^{-1})	Insensitivity factor
SSK ($n = 8$)	$2.5 \times 10^5 \pm 1.3 \times 10^5$	
22S84 ($n = 6$)	$9.8 \times 10^4 \pm 1.3 \times 10^4$	2.5
6005 ($n = 6$)	$1.7 \times 10^5 \pm 0.5 \times 10^5$	1.4
6002 ($n = 12$)	$4.4 \times 10^4 \pm 2.2 \times 10^4$	5.5

Insensitivity factor is expressed as the ratio of bimolecular rate constants k_i SSK/k_i field sample

k_i values are expressed as the mean of several independent assays using at least two different diethyl paraoxon concentrations each. Number of independent assays is represented as (*n*=)

value observed for the 6002 sample led to a 5.5-fold difference in the AChE insensitivity of this isolate toward paraoxon compared to AChE sensitivity in SSK strain. This was associated with two to sixfold increase of the time required to obtain 50% of AChE inhibition compared to SSK strain depending on the inhibitor concentration.

AChE inhibition screening

IC_{50} values obtained from the SSK strain were $1.52 \times 10^{-7} \pm 0.18$ M. IC_{50} values obtained from field samples ranged between 1.29×10^{-7} and 1.47×10^{-6} M. The insensitivity factor, calculated as a ratio between IC_{50} values obtained from field samples with that with the SSK strain, both obtained during a single assay, are represented in Fig. 2. The 39 field samples harbored a range in the IC_{50} ratio compared to that of SSK strain between 1 and 6. Over the 39 analyzed samples, none provided a IC_{50} ratio over 10.

Discussion

Validation of the tests

The results reported in the present study showed that the selected method is convenient and sensitive enough to detect subtle changes in AChE activities between field populations. AChE activity was linear up to 200 protonymphs per replicate and 40 min of incubation (Fig. 1a, b). Hydrolysis of ASCh appears to be catalytically mediated by AChE because of a strong decrease in activity when protonymphs extracts were incubated with 500 μM eserine or after thermal inactivation (Fig. 1c).

Kinetics

The kinetic properties of *D. gallinae* AChE in the reference strain SSK for ASCh were characterized and michaelian parameters were compared to those obtained from field samples. K_M values were comparable to that obtained from mites such as *Tetranychus urticae* Koch (Tsagkarakou et al. 2002). It has been reported that incubations performed with Triton X-100 may significantly affect the kinetic constant K_M exhibiting a competitive

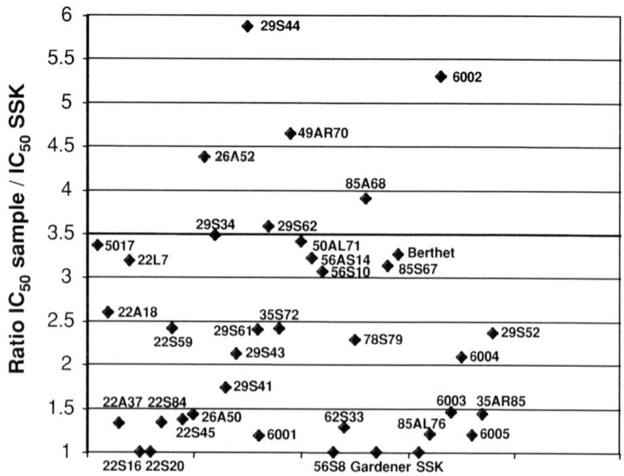

Fig. 2 Acetylcholinesterase inhibition studies of 39 field-collected samples. IC_{50} for acetylcholinesterase inhibition by diethyl paraoxon were determined according to "Material and methods". Each data point represents the ratio of IC_{50} value of the corresponding field-collected sample divided by IC_{50} value of the sensitive reference strain (SSK) obtained in the same assay

inhibition (Chen et al. 2001; Rosenfeld et al. 2001). Nevertheless, we were unable to investigate the kinetic properties of the native enzyme because of a lack of efficiency in solubilizing AChE enzyme when the detergent was omitted from the lysis buffer.

All tested field samples exhibited similar K_M values and different V_{max} values (some field isolates with increased V_{max} values). It has been reported that insensitive AChE in *Boophilus microplus* exhibited a reduced K_M value relative to the susceptible enzyme associated with a corresponding lower V_{max} (Nolan and Schnitzerling 1975). Smissaert found similar results in *T. urticae* sensitive and resistant strains (Smissaert 1964). Conversely in the latter species, lower K_M values in AChE sensitive strain associated with similar V_{max} values as compared to resistant strains were reported (Stumpf et al. 2001). These modifications in kinetics parameters were considered to be the consequence of structural changes in the enzyme and are the biochemical support of the severe fitness cost that has been observed in most populations with insensitive AChE that expressed a reduction of AChE activity in synapses (Lenormand et al. 1999). In our study, the reference SSK strain and the field sample (6001) exhibited a significant lower V_{max} value compared to the other samples. Although low AChE activity is a characteristic of a resistant phenotype (Lee and Bantham 1966), none of AChE inhibition studies conducted with extracts from these two populations exhibited specific characteristics of a resistant strain (Fig. 2; Table 2). Interestingly, comparisons of the kinetic parameter V_{max} between the different populations clearly showed that polymorphism in AChE expression exists among isolates of *D. gallinae* under test. Whether it can be supposed that a strain which has been maintained in laboratory culture for several years without any acaricide treatment may led to produce an homogenous population with low intrinsic AChE activity, it is surprising to detect a field-collected population harboring the same low specific activity without any apparent disadvantage.

As AChE inhibition by diethyl paraoxon follows a first-order kinetic, the bimolecular rate constant was determined for SSK and three field isolates as this parameter appears to be a much better index than the usual—but less time and mites consuming—IC_{50} (Aldridge and Davison 1952). The k_i values obtained in *D. gallinae* were comparable to those described in *T. urticae* (Tsagkarakou et al. 2002), but differences between *D. gallinae* isolates remained small compared to those observed between sensitive and experimentally selected OP resistant strain of *T. urticae* (k_i differences of 39 fold in Tsagkarakou et al. 2002). Indeed, kinetic analysis of the interaction of AChE from these three different field isolates (22S84-6005-6002) with diethyl paraoxon revealed a maximum difference between the k_i values of ~5–6 in the favor of the 6002 isolate (Table 2; Fig. 3) and was associated with a higher IC_{50} value (Fig. 2). The kinetics of AChE clearly showed that the enzyme, in this isolate, was different than those measured in other *D. gallinae* isolates. Nevertheless, whether both the lower k_i value and the higher IC_{50} value evoked target site insensitivity, it is worth to note that the insensitivity factor, obtained in the 6002 isolate, remained low compared to factors observed in experimentally confirmed AChE-resistant strains in other related mites (Tsagkarakou et al. 2002). The decrease in k_i value in this particular isolate should probably be interpreted as a probable decrease in AChE affinity to diethyl paraoxon even if a modification of the phosphorylation rate should not be excluded as it was shown in OP resistant *B. microplus* (Pruett 2002). In this species, the bimolecular rate constant was most affected by a slower rate of enzyme phosphorylation. At last, kinetic studies of AChE in field samples of *D. gallinae* exhibited a moderate heterogeneity in these activities that may be associated with different sensitivities to OP. The biological significance of this polymorphism and its potential impact on the control of the Poultry Red Mite in farms remains to be evaluated. In order to get an overview of this polymorphism prevalence, 39 populations from field isolates were screened and their IC_{50} measured.

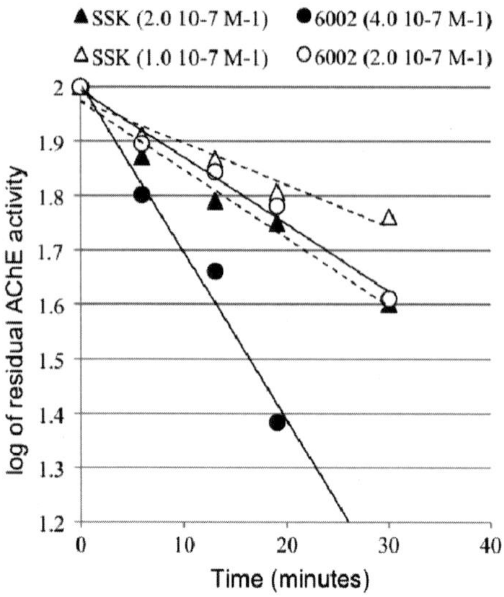

Fig. 3 Graph representative of the evolution of residual AChE activity from isolate 6002 and strain SSK with different incubation times and two different diethyl paraoxon concentrations. Isolate 6002 and strain SSK have been incubated with two different sets of concentrations of diethyl paraoxon (SSK: 2.0×10^{-7} and 1.0×10^{-7} M^{-1}; 6002: 4.0×10^{-7} and 2.0×10^{-7} M^{-1}). As inhibition is stopped by the introduction of substrate ASCh, the difference of incubation time generates different levels of inhibition with a single OP concentration. The apparent bimolecular rate k_i was extracted from the regression of each slope

Screening of 39 field populations

Around 10% of populations under test exhibited a maximum 6-fold difference in the inhibitory effect of diethyl paraoxon compared to SSK strain (Fig. 2) whereas a 100-fold difference in the inhibitory effect of diethyl paraoxon is classically observed in tetranychid pest species (Stumpf et al. 2001) even if lower insensitivity factors have sometimes been reported for other OP compounds such as dichlorvos (ratios ranged between 25 and 38, Zahavi and Tahori 1970). Insensitivity factors from these 39 isolates were not as high as compared to other acari proved to be resistant. Indeed, even if AChE OP-insensitivity is defined by a slower rate of AChE inhibition in the resistant phenotype and that the insensitivity factor can vary depending on the species under test or the inhibitor compound used, insensitivity factors (IC_{50} ratios between a population under test and SSK from a single assay) obtained in our results appear too low to consider any field-collected population as resistant.

However, results of this study clearly showed that AChE from field-collected populations of *D. gallinae* exhibited different susceptibilities to diethyl paraoxon. But no target resistance has been detected in isolates under test.

Conclusion

This study provides the basis for the development of diagnostic tools that can be used for management of possible AChE resistance in *D. gallinae* against OPs and carbamates insecticides. A screening on 39 field samples has revealed moderate differences in AChE sensitivity to paraoxon between field populations but AChE insensitivity to OPs has been considered to be too weak in comparison with analyses on other Arthropoda in literature. Thus, in spite of the existence of selection pressure, no important AChE insensitivity has occurred in *D. gallinae*. But, our results clearly show that isolates which were sampled in independent farms revealed distinct inhibition kinetics suggesting the existence of AChE polymorphism in *D. gallinae*. Additionally, it would be interested to test whether the polymorphism detected in field isolates in present study is selectable under laboratory conditions under elevated OP pressure.

Acknowledgments We are grateful for having provided SSK strain to O. Kilpinen and N. Hansen (Danish Institute of Agricultural Sciences, Denmark) and for their help and advices to A. Micoud, C. Brazier and C. Mottet (Service Régional de la Protection des Végétaux, France). We also want to thank Mehdi Gharbi (IUT, Villeurbanne, France), Jennifer Lanneau (IUT, Villeurbanne, France) and Coralie Pulido (LGTA, Saint-Genis Laval, France) for their technical help.

References

Aldridge WN (1950) Some properties of specific cholinestarase with particular reference to the mechanism of inhibition by diethyl *p*-nitrophenyl triphosphate (E605) and analogues. Biochem J 46:451–460

Aldridge WN, Davison AN (1952) The inhibition of erythrocyte cholinesterase by tri-esters of phosphoric acid. Biochem J 52:62–70

Baxter GD, Green P et al (1999) Detecting resistance to organophosphates and carbamates in the cattle tick *Boophilus microplus*, with a propoxur-based biochemical test. Exp Appl Acarol 23:907–914. doi: 10.1023/A:1006364816302

Chauve C (1998) The poultry red mite *Dermanyssus gallinae* (De Geer, 1778): current situation and future prospects for control. Vet Parasitol 79:239–245. doi:10.1016/S0304-4017(98)00167-8

Chen Z, Newcomb R, Forbes E et al (2001) The acetylcholinesterase gene and organophosphorus resistance in the Australian sheep blowfly, *Lucilia cuprina*. Insect Biochem Mol Biol 31:805–816. doi: 10.1016/S0965-1748(00)00186-7

Ellman GL, Courtney KD et al (1961) A new, rapid colorimetric determination of acetylcholinesterase activity. Biochem Pharmacol 7:88–95. doi:10.1016/0006-2952(61)90145-9

Fournier D, Mutero A (1994) Modification of the acetylcholinesterase as a mechanism of resistance to insecticides. Comp Biochem Physiol 108C:19–31

Keçeci T, Handemir E, Orhan G (2004) The effect of *Dermanyssus gallinae* infestation on hematological values and body weights of cocks. Turkiye Parazitol Derg 28:192–196

Keita A, Pagot E, Pommier P et al (2006) Efficacy of phoxim 50% E.C. (ByeMite) for treatment of *Dermanyssus gallinae* in laying hens under field conditions. Rev Med Vet 157:590–594

Kirkwood AC (1967) Anaemia in poultry infested with the red mite *Dermanyssus gallinae*. Vet Rec 80(17):514–516

Lee RM, Bantham P (1966) The activity and organophosphate inhibition of cholinesterase from susceptible and resistant ticks (Acari). Entomol Exp Appl 9:13–24. doi:10.1007/BF00341156

Lenormand T, Bourguet D, Guillemaud T et al (1999) Tracking the evolution of insecticide resistance in the mosquito *Culex pipiens*. Nature 400:861–864. doi:10.1038/23685

Nabeshima T, Mori A, Kozaki T et al (2004) An amino acid substitution attributable to insecticide-insensitivity of acetylcholinesterase in a Japanese encephalitis vector mosquito, *Culex tritaeniorhynchus*. Biochem Biophys Res Commun 313:794–801. doi:10.1016/j.bbrc.2003.11.141

Newcomb RD, Campbell PM, Russel RJ et al (1997) cDNA cloning, Baculovirus-expression and kinetic properties of the esterase, E3, involved in organophosphorus resistance in *Lucilia cuprina*. Insect Biochem Mol Biol 27:15–25. doi:10.1016/S0965-1748(96)00065-3

Nolan J, Schnitzerling HJ (1975) Characterisation of acetylcholinesterases of acaricide resistant and susceptible strains of the cattle tick *Boophilus microplus*. I. Extraction of the critical component and comparison with enzyme from other sources. Pest Biochem Physiol 5:178–188

Nordenfors H, Höglund J, Tauson R et al (2001) Effect of permethrin impregnated plastic strips on *Dermanyssus gallinae* in loose-housing systems for laying hens. Vet Parasitol 102:121–131. doi: 10.1016/S0304-4017(01)00528-3

Pruett JH (2002) Comparative inhibition kinetics for acetylcholinesterases extracted from organophosphate resistant and susceptible strains of *Boophilus microplus* (Acari:Ixodidae). J Econ Entomol 95:1239–1244

Reuveny H, Cohen E (2004) Evaluation of mechanisms of azinphos-methyl resistance in the codling moth *Cydia pomonella* (L.). Insect Biochem Physiol 57:92–100. doi:10.1002/arch.20016

Rosenfeld C, Kousba A, Sultatos LG (2001) Interactions of rat brain acetylcholinesterase with the detergent Triton X-100 and the organophosphate paraoxon. Toxicol Sci 63:208–213. doi:10.1093/toxsci/63.2.208

Smissaert HR (1964) Cholinesterase inhibition in spider mites susceptible and resistant to organophosphates. Science 143:129–134. doi:10.1126/science.143.3602.129

Stone BF, Nolan J, Schuntner CA (1976) Biochemical genetics of resistance to organophosphorus acaricides in three strains of the cattle tick, *Boophilus microplus*. Aust J Biol Sci 29:265–279

Stumpf N, Zebitz C, Kraus W et al (2001) Resistance to organophosphates and biochemical genotyping of acetylcholinesterases in *Tetranychus urticae* (Acari: Tetranychidae). Pestic Biochem Physiol 69:131–142. doi:10.1006/pest.2000.2516

Tsagkarakou A, Pasteur N, Cuany A et al (2002) Mechanisms of resistance to organophosphates in *Tetranychus urticae* (Acari: Tetranychidae) from Greece. Insect Biochem Mol Biol 32:417–424. doi:10.1016/S0965-1748(01)00118-7

Valiente Moro C, Chauve C, Zenner L (2005) Vectorial role of some dermanyssoid mites (Acari, Mesostigmata, Dermanyssoidea). Parasite 12:99–109

Valiente Moro C, Fravalo P, Amelot M et al (2007) Colonization and invasion in chicks experimentally infected with *Dermanyssus gallinae* contaminated by *Salmonella enteritidis*. Avian Pathol 36:307–311. doi:10.1080/03079450701460484

Wang JJ, Cheng WX et al (2004) The effect of the insecticide dichlorvos on esterase extracted from the psocids, *Liposcelis bostrychophila* and *L. entomophila*. J Insect Sci 4:23–27

Weill M, Lutfalla G, Mogensen K et al (2003) Comparative genomics: Insecticide resistance in mosquito vectors. Nature 423:136–137. doi:10.1038/423136b

Wood HP (1917) The chicken mite: its life history and habits United States Department of Agriculture, Washington, DC. Bull 553:1–14

Xu G, Fang QQ, Keirans JE et al (2003) Cloning and sequencing of putative acetylcholinesterase cDNAs from the American dog tick, *Dermacentor variabilis,* and the brown dog tick, *Rhipicephalus sanguineus* (Acar : Ixodidae). J Med Ent 40:890–896

Zahavi M, Tahori AS (1970) Sensitivity of acetylcholinesterase in spider mites to organophosphorous compounds. Biochem Pharmacol 19:219–225. doi:10.1016/0006-2952(70)90342-4

Zeman P, Železný J (1985) The susceptibility of the poultry red mite, *Dermanyssus gallinae* (De Geer, 1778), to some acaricides under laboratory conditions. Exp Appl Acarol 1:17–22. doi:10.1007/F01262196

Zhu KY, Lee SH, Clarck JM (1996) A point mutation of acetylcholinesterase associated with azinphosmethyl resistance and reduced fitness in colorado potato beetle. Pestic Biochem Physiol 55:100–108. doi:10.1006/pest.1996.0039

In vitro efficacies of oils, silicas and plant preparations against the poultry red mite *Dermanyssus gallinae*

Veronika Maurer · Erika Perler · Felix Heckendorn

Originally published in the journal Experimental and Applied Acarology, Volume 48, Nos 1–2, 31–41.
DOI: 10.1007/s10493-009-9254-2

Abstract The aim of this study was to test the effectiveness of physically acting substances (oils and silicas) and plant preparations for the control of the poultry red mite *Dermanyssus gallinae* (De Geer 1778). Reproduction and survival of fed *D. gallinae* females were evaluated in vitro for a total of 168 h using the "area under the survival curve" (AUC) to compare survival of the mites between treatments. Four oils (two plant oils, one petroleum spray oil and diesel), one soap, three silicas (one synthetic amorphous silica, one diatomaceous earth (DE) and one DE with 2% pyrethrum extract) and seven plant preparations (derived from *Chrysanthemum cineariaefolium*, *Allium sativum*, *Tanacetum vulgare*, *Yucca schidigera*, *Quillaja saponaria*, *Dryopteris filix-mas*, and *Thuja occidentalis*) were tested at various concentrations. All the oils, diesel and soap significantly reduced *D. gallinae* survival. All silicas tested inhibited reproduction. DE significantly reduced mite survival, but amorphous silica was less effective in vitro. Except for pure *A. sativum* juice and the highest concentration of *C. cineariaefolium* extract, the plant preparations tested resulted in statistically insignificant control of *D. gallinae*.

Keywords *Dermanyssus gallinae* · Ectoparasite · Control · Silicas · Oils · Plant extracts

Introduction

The poultry red mite *Dermanyssus gallinae* (De Geer 1778) is regarded as the most important ectoparasite of laying hens in organic as well as conventional egg production in Europe (Maurer et al. 1993; Höglund et al. 1995; Fiddes et al. 2005). The haematophagous mite is a nocturnal feeder and spends the daylight hours in refugia in the vicinity of the hens. At high population densities *D. gallinae* can cause severe anaemia and associated

V. Maurer (✉) · E. Perler · F. Heckendorn
Research Institute of Organic Agriculture (FiBL), Ackerstrasse, 5070 Frick, Switzerland
e-mail: veronika.maurer@fibl.org

mortality (Kirkwood 1967). Even low mite populations can irritate hens to the extent that they refuse to use the henhouse or rest on the perches (Maurer et al. 1995; Kilpinen et al. 2005). Production can also be affected by reductions in egg production and egg quality, where mites may cause staining of the egg shell surface (Cencek 2003). Since *D. gallinae* may also act as a vector for numerous pathogens of medical and veterinary importance, spread of disease is another problem associated with this mite in poultry systems (Chirico et al. 2003). *D. gallinae* causes nuisance and dermatitis in people working in heavily infested poultry houses (Rosen et al. 2002).

Dermanyssus gallinae are typically controlled by treating the poultry house installations rather than the hens themselves (Hoop 2008). Several synthetic acaricide classes are widely used for mite control (organophosphates, pyrethroids, carbamates), but *D. gallinae* has developed resistance against some of these compounds (Zeman and Zelezny 1985; Beugnet et al. 1997; Nordenfors et al. 2001). In addition, some compounds are unsuitable for food safety and environmental reasons (Chauve 1998). On organic farms, synthetic acaricides may be used as a last resort, but *D. gallinae* control should primarily be achieved by preventive measures and acaricides of natural origin according to national and international regulations (e.g. the Council Regulation (EC) No 834/2007; EC 2007). A three-stage control system is widely applied on Swiss organic farms. The concept includes management practices as a first stage (such as cleaning and disinfection of the empty house after each cycle). As a second stage, physically acting substances (such as oil and desiccant dusts) are used during flocks. As a third and final stage, acaricides of natural origin are selectively applied to highly infested sites in the house. Physically acting substances such as oils and dusts offer an attractive alternative to synthetic acaricides because resistance is less likely to occur. Acaricides of natural origin have been researched for several areas of pest management with encouraging results (Isman 2006). *Varroa destructor* and the tracheal mite *Acarapis woodi* are examples for parasitic mite species which can successfully be controlled with acaricides of natural origin (Rice et al. 2002). Some work has already been conducted with natural plant preparations for their effect on *D. gallinae* where recent studies have focused on in vitro effects of oriental medicinal plant extracts (Kim et al. 2007) and plant essential oils (Kim et al. 2004; George et al. 2008a, b).

With both natural plant preparations and physically acting substances showing promise for *D. gallinae* management, it is important that comparisons are made between these methods to maximize the effectiveness of existing or potential non-synthetic *D. gallinae* control strategies. This paper deals with the in vitro effectiveness of oils, silicas, and selected plant preparations for the control of *D. gallinae* with the aim of comparing the efficacy of these different non-synthetic control options.

Materials and methods

Dermanyssus gallinae

Fed *D. gallinae* females were obtained from a naturally infested poultry house at the Research Institute of Organic Agriculture, Ackerstrasse, Switzerland. The mites were collected in traps consisting of a u-shaped aluminium-profile containing a strip of fabric in a zigzag fold fixed under the perches during one night (Maurer et al. 1993). All mites were used for the tests within 1 day of collection.

Test substances

Table 1 gives an overview of the test substances, their origin, and concentration tested. A total of four oils (two plant oils, one petroleum spray oil and diesel), one soap, three silicas (one synthetic amorphous silica, one diatomaceous earth (DE) and one DE with 2% pyrethrum extract) and plant preparations based on *Chrysanthemum cineariaefolium*, *Allium sativum*, *Tanacetum vulgare*, *Yucca schidigera*, *Quillaja saponaria*, *Dryopteris filix-mas*, and *Thuja occidentalis* were tested. Six of these test substances were tested at various concentrations and/or mixtures.

In vitro assay

Oils and plant preparations

Five fed adult female mites were transferred into plastic vials (ø 33 × 16 mm) with a tightly closing lid containing a filter paper disk (ø 27 mm) saturated with the test substance. The following controls were used: untreated (water) for water extracts and alcohol (EtOH 10%) for EtOH extracts.

Silicas (powders and liquid formulation)

Five fed adult female mites were transferred into the same type of plastic vials containing 0.05 or 0.005 g of the powders (9 or 0.9 mg/cm^2) or a filter disk treated with the amount of liquid test product containing 0.05 g of dry matter and air dried before the test. Untreated vials served as controls.

Numbers of replicates used per concentration are indicated in Table 1. Vials were placed together to assure a similar macro-environment. Treatments were randomly assigned to the vials. The mites were kept at 27°C in dark during the experiments. Surviving mites were counted after 4, 24, and 168 h under a binocular microscope. Mites showing no symptoms as well as stricken mites (movements, but no locomotion) were classified as "alive". Mites were considered "dead" if no movement was visible even after a gentle touch with a paint brush. The presence of offspring (eggs, larvae and/or unfed protonymphs) was qualitatively recorded at each observation interval.

Data analysis

For statistical comparison of *D. gallinae* survival, the integral of the survival curves was estimated for each vial (trapezoidal integration). The calculated "area under the curve" (AUC) has units of "percent-hours" (Campbell and Madden 1990). For each vial, AUCs were calculated for the periods from 0 to 4 h and from 0 to 168 h after treatment, respectively (herein after referred to as AUC4 and AUC168). The non parametric Kruskal–Wallis test was used to provide estimates of the global difference between groups separately for: oils (nine treatment groups including the control), silicas (five treatment groups), water-based plant preparations (13 treatment groups) and ethanol-based plant preparations (three treatment groups). In case of a significant result of the global test, pair-wise comparisons were made between the treatment groups using the Mann–Whitney rank sum test. *P*-values of the pair-wise comparisons where adjusted for multiple comparisons according to the formula:

$$p\text{ - value} = \alpha \times 2/k(k-1)$$

Table 1 Test substances used in in vitro assay with *Dermanyssus gallinae*: origins, concentrations and numbers of replicates

Scientific name	Common name	Plant part/extraction method	Supplier	Specification/brand name of commercial product	Active ingredient in commercial product (g/l or g/kg)	Concentrations tested (g/l)	Replicates (number of vials)
Oils/soap							
H_2O	Water, control					1,000	10
	Rapeseed oil		Local shop	Food-grade	1,000	1,000	10
	Orange oil		Wigger, Althäusern (CH)	Parasitex®	50	50, 5, 2.5	10, 15, 10
	Diesel (fuel)		Local shop	Fuel	1,000	1,000	15
	Petroleum spray oil		Omya, Oftringen (CH)	Mineral Oil Omya®	990	990	10
	Soap		Biocontrol, Grossdietwil (CH)	Natural®	1,000	1,000, 100, 10	10
Silicas							
	Empty vial, control					0	20
	Synthetic amorphous silica		Evonik Degussa, Frankfurt (DE)	Indispron D110®	1,000	0.005[a]	10
	Diatomaceous earth		Biovet, Grossdietwil (CH)	Gallo-Sec®	1,000	0.005, 0.0005[a]	10
	Diatomaceous earth & pyrethrum extract		Agro-Hygiene AG, Wald (CH)	FLY-END Acaricide powder®	9,980 & 20	0.0005[a]	10
Plant preparations							
Commercial product, water extracts and mechanically pressed juices							
H_2O	Water, control					1,000	50
Chrysanthemum cinerariaefolium	Pyrethrum		Biovet, Grossdietwil (CH)	Pyri-Fly®	20	2, 0.2, 0.02	10
Allium sativum	Garlic	Bulb/mechanically pressed fresh juice	Local shop	Organic production		1,000, 100, 10	10, 5, 5
Tanacetum vulgare	Tansy	Superficial parts; start of full bloom/water extract	In house production			1,000	10

Table 1 continued

Scientific name	Common name	Plant part/extraction method	Supplier	Specification/brand name of commercial product	Active ingredient in commercial product (g/l or g/kg)	Concentrations tested (g/l)	Replicates (number of vials)
Yucca schidigera	Mohave yucca	Leaves/mechanically pressed, thermally condensed	Desert King International, San Diego (USA)			1,000, 100	10, 10
Quillaja saponaria	Soapbark tree	Bark/water extract	Desert King International, San Diego (USA)			1,000, 100	20, 10
Yucca:Quillaja 1:1	See above	See above	Desert King International, San Diego (USA)			500 : 500	10
Ethanol extracts							
EtOH	Ethanol, control				960	100	10
Dryopteris filix-mas	Male-fern	Leaves/ethanol extract	In house production			100	10
Thuja occidentalis	Arborvitae	Leaves/ethanol extract	In house production			100	10

[a] g/vial

(Bonferroni Correction; Lu and Fang 2003), where k = number of comparisons and α = agreed chance of falsely positive result (here 0.05). Reproduction data was analysed separately for oils, silicas and water-based plant preparations using logistic regression models. All data was analysed using STATA® 9.0 (StataCorp LP, 4905 Lakeway Drive, TX 77845, USA) software.

Results

Table 2 presents the proportion of vials with reproduction and the AUC of the oil, silica, and plant preparation treatments. In all control vials reproduction occurred during the experiment and AUCs were close to or at the maximum possible values.

Oils

Plant (rapeseed and orange) as well as petroleum spray oil and diesel reduced *D. gallinae* reproduction and the AUCs, including AUC4, which indicates a rapid effect of those products. Treatments with the petroleum spray oil and diesel significantly reduced the AUC168 values by 95% and more. Orange oil was significantly effective at concentrations of 5% only. Mites treated with 100% soap did not reproduce, but the effect on survival was significant only after 1 week (AUC168).

Silicas

Female mites did not produce eggs in any of the silica treatments. DE with or without additional pyrethrum increased mortality of the mites later than 4 h after treatment, reflected by high AUC4 and low AUC168. Synthetic amorphous silica was not effective.

Plant preparations

Egg production was observed in all treatments except for the highest dose of pyrethrum and the higher doses of garlic juice. Pure garlic juice was the only plant preparation which quickly killed *D. gallinae*, as reflected by a significantly reduced AUC4 by 50%. Of the several preparations tested, only pyrethrum 0.2 and 0.02% and garlic juice 10 and 100% significantly reduced AUC168 by more than 50%. A dose-response of the pyrethrum treatment was seen in the % vials with reproduction as well as in the AUC168 values.

Discussion

Untreated mites reproduced and their survival was close to the maximum possible value. This indicates that the experimental conditions used were favourable for survival and reproduction of the fed *D. gallinae* females. George et al. (2008b) suggest that *D. gallinae* are more susceptible to the effects of plant essential oils after starving for 3 weeks than recently fed mites. Our experimental conditions using fed females therefore represent a

Table 2 Area under the curve (AUC) of *Dermanyssus gallinae* 4 h (AUC4) and 7 days (AUC168) after treatment with potential acaricides

Name (brand name or specification)	Concentration (g/l)	Reproduction		AUC4		AUC168	
		Proportion of replicates with juveniles	Difference to control (%)	Mean (percent-hours[b])	Difference to control (%)	Mean (percent-hours[b])	Difference to control (%)
Oils/soap							
Control	–	1		400		16,470	
Rapeseed oil	1,000	0.4	−60*	284	−29.0*	2,632	−84.1*
Orange oil	50	0	−100**	256	−36.0*	2,012	−87.8*
Orange oil	5	0.66	−34	336	−16.0	8,021	−51.4
Orange oil	2.5	1	0	392	−2.0	13,976	−15.4
Diesel	1,000	0	−100**	205	−48.7*	232	−98.6*
Petroleum spray oil	990	0	−100**	304	−24.0	824	−95.0*
Soap	1,000	0	−100**	324	−19.0	3,568	−78.4*
Soap	100	0.7	−30	344	−14.0	7,880	−52.3
Silicas							
Control (empty vial)	–	1		400		16,152	
Synthetic amorphous silica	0.005[a]	0	−100**	400	0.0	7,796	−51.7
Diatomaceous earth	0.005[a]	0	−100**	352	−12.0	1,112	−93.1*
Diatomaceous earth	0.0005[a]	0	−100**	352	−12.0*	1,194	−92.6*
Diatomaceous earth & Pyrethrum extract	0.0005[a]	0	−100**	400	0.0	1,892	−88.3*
Plant preparations							
commercial product, water extracts and mechanically pressed juices							
Control (water)	–	1		400		16,470	
Chrysanthemum cinerariaefolium	2	0	−100**	400	0.0	1,400	−91.5*
Chrysanthemum cinerariaefolium	0.2	0.33	−67*	346	−13.5	5,581	−66.1*
Chrysanthemum cinerariaefolium	0.02	1	0	400	0.0	1,450	−11.7
Allium sativum	1,000	0	−100**	200	−50.0*	200	−98.8*

Table 2 continued

Name (brand name or specification)	Concentration (g/l)	Reproduction		AUC4		AUC168	
		Proportion of replicates with juveniles	Difference to control (%)	Mean (percent-hours[b])	Difference to control (%)	Mean (percent-hours[b])	Difference to control (%)
Allium sativum	100	0	−100**	400	0.0	5,008	−69.6*
Allium sativum	10	1	0	400	0.0	15,936	−3.2
Tanacetum vulgare	1,000	1	0	400	0.0	15,504	−5.9
Yucca schidigera	1,000	1	0	400	0.0	11,432	−30.6*
Yucca schidigera	100	1	0	400	0.0	14,352	−12.9
Quillaja saponaria	1,000	1	0	392	−1.2	1,244	−24.4*
Quillaja saponaria	100	1	0	400	0.0	13,056	−20.7
Yucca:Quillaja 1:1	1,000	1	0	400	0.0	10,320	−37.3*
Ethanol extracts							
Control (EtOH 10%)	100	1	0	400		16,512	
Dryopteris filix-mas	100	1	0	400	0.0	14,456	−12.5
Thuja occidentalis	100	1	0	400	0.0	16,665	+0.9

[a] g/vial

[b] Campbell and Madden (1990)

* $P < 0.05$; ** $P < 0.01$

more severe test of product efficacy than tests with starved mites as proposed e.g. by Thind and Ford (2006).

The effects of oils on plant pest insects and mites have been investigated in some detail (e.g. Agnello et al. 2003; Fernandez et al. 2005). Mineral oil was more toxic to adult phytoseiid mites than plant oil (Momen et al. 2006). A study on the effects of oils on *D. gallinae* showed that a mineral oil developed for agricultural use (OPPA) caused 100% mite mortality after 2 h of exposure when sprayed directly on the mites (Guimaraes and Tucci 1992). None of the oils tested in the current study acted as quickly and completely, probably because the contact of the mites with the test substances in the experimental setup was reduced as compared to that in the aforementioned study, where mites where completely covered by means of spraying. Diesel reduced the AUC4 by 50% in the current study, and the AUC168 by almost 100%. However, odour and associated risk of egg contamination are a serious drawback of diesel, which is not recommended for use in poultry houses on this basis (Hoop 2008). Slightly higher AUC168s were attained by the odourless and relatively cheap petroleum spray oil and rapeseed oil, making these interesting alternatives to diesel. The effects of the undiluted product containing 5% orange oil were similar to those of diesel and petroleum spray oil. The disadvantage of this treatment is the high cost of the effective concentration.

Inert dusts based on silica are commonly used as desiccating agents against stored-product pests (Collins 2006; Palyvos et al. 2006). Silicas act physically and their activity is not dependent on metabolic pathways. In our experiment, neither the treatment dose (mg/vial) nor the addition of Pyrethrum extract improved upon the already favourable efficiency of DE. Mortality from DE exposure is mainly a result of desiccation (Saez and Fuentes Mora 2007), and arthropods are therefore not expected to develop genetic resistance. However, insects have been shown to develop behavioural responses to avoid contact with such products (Ebeling 1971), and this may also be the case for mites. Drawbacks of DE are the formation of dust during treatment and the decrease of efficacy due to high humidity. Therefore, substantial efforts have been put into the development of liquid formulations (Lamina and Kruner 1966) such as the synthetic amorphous silica used in the present study. In the in vitro setup used in the current work, the liquid formulation completely suppressed reproduction of *D. gallinae* females, but AUC168 was not significantly different from the control. This result suggests insufficient efficiency of the amorphous silica compared to DE. However, field experiments performed by Maurer and Perler (2006) in heavily infested layer houses revealed a longer residual effect of the liquid amorphous silica compared to DE.

Garlic is often used in folk medicine, but scientific studies on the effects of *A. sativum* on mites are scarce. A study by Birrenkott et al. (2000) showed that repeated topical applications of garlic juice (10%) on hens heavily infested with the Northern fowl mite *Ornithonyssus sylviarum* significantly reduced the level of infestation. The authors suggest that this was mainly due to a repellent effect preventing re-infestation and not to direct acaricidal effects. In our in vitro test, direct effects on oviposition and on survival of the closely related species *D. gallinae* have been demonstrated. This indicates that fresh garlic juice can have both, direct and indirect effects on gamasid mites. From the literature, it remains unclear whether garlic oil has the same insecticidal properties as fresh juice. A study of Amonkar and Reeves (1970) demonstrated that partly purified garlic oil had a higher toxic effect on mosquito larvae than the crude extract. In contrast, a 10% concentrate made from commercial chopped garlic provided control of whiteflies, but commercial garlic oil gave little or no control in an experiment by Flint et al. (1995).

Before garlic can be considered as a valuable component of a strategy against *D. gallinae*, the question of conservation and standardisation of the crude extract has to be solved.

Except for pure garlic juice and the highest concentration of pyrethrum extract, the plant preparations tested in our study resulted in <70% or no significant reduction of *D. gallinae*. Therefore, and also supported by observations of others (George et al. 2008b), it may be more suitable to use plant preparations as a treatment for starved mites in the poultry houses between layer flocks, rather than to apply them during flocks when mites have the opportunity to recently feed. In contrast, our experiment showed that physically acting products such as rapeseed oil, petroleum spray oil, soap or DE were effective on fed mites. These products therefore seem better suited for application in cases of severe *D. gallinae* infestations during flocks.

Acknowledgments The authors gratefully acknowledge funding from the European Community financial participation under the Sixth Framework Programme for Research, Technological Development and Demonstration Activities, for the Integrated Project QUALITYLOWINPUTFOOD, FP6-FOOD-CT-2003-506358. The views expressed in this publication are the sole responsibility of the author(s) and do not necessarily reflect the views of the European Commission. Neither the European Commission nor any person acting on behalf of the Commission is responsible for the use which might be made of the information contained herein. Mention of a proprietary chemical or a trade name does not constitute a recommendation or endorsement.

References

Agnello AM, Reissig WH, Kovach J, Nyrop JP (2003) Integrated apple pest management in New York State using predatory mites and selective pesticides. Agric Ecosyst Environ 94:183–195. doi:10.1016/S0167-8809(02)00026-9

Amonkar SV, Reeves EL (1970) Mosquito control with active principle of garlic, *Allium sativum*. J Econ Entomol 63:1172–1175

Beugnet F, Chauve C, Gauthey M, Beert L (1997) Resistance of the red poultry mite to pyrethroids in France. Vet Rec 140:577–579

Birrenkott GP, Brockenfelt GE, Greer JA, Owens MD (2000) Topical application of garlic reduces northern fowl mite infestation in laying hens. Poult Sci 79:1575–1577

Campbell CL, Madden LV (1990) Introduction to plant disease epidemiology. Wiley, New York, p 532

Cencek T (2003) Prevalence of *Dermanyssus gallinae* in poultry farms in silesia region in Poland. Bull Vet Inst Pulawy 47:465–469

Chauve C (1998) The poultry red mite *Dermanyssus gallinae* (De Geer, 1778): current situation and future prospects for control. Vet Parasitol 79:239–245. doi:10.1016/S0304-4017(98)00167-8

Chirico J, Eriksson H, Fossum O, Jansson D (2003) The poultry red mite, *Dermanyssus gallinae*, a potential vector of *Erysipelothrix rhusiopathiae* causing erysipelas in hens. Med Vet Entomol 17:232–234. doi:10.1046/j.1365-2915.2003.00428.x

Collins DA (2006) A review of alternatives to organophosphorus compounds for the control of storage mites. J Stored Prod Res 42:395–426. doi:10.1016/j.jspr.2005.08.001

Ebeling W (1971) Sorptive dusts for pest control. Annu Rev Pest Control 16:123–158

EC (2007) Council Regulation (EC) No 834/2007 of 28 June 2007 on organic production and labelling of organic products and repealing. Regulation (EEC) No 2092/91. Official Journal of the European Communities, L189/1 (20.7.2007), pp 1–23

Fernandez DE, Beers EH, Brunner JF, Doerr MD, Dunley JE (2005) Effects of seasonal mineral oil applications on the pest and natural enemy complexes of apple. J Econ Entomol 98:1630–1640

Fiddes MD, Le Gresley S, Parsons DG, Epe C, Coles GC, Stafford KA (2005) Prevalence of the poultry red mite (*Dermanyssus gallinae*) in England. Vet Rec 157:233–235

Flint HM, Parks NJ, Holmes JE, Jones JA, Higuera CM (1995) Tests of garlic oil for control of the silverleaf whitefly, *Bemisia argentifolii* bellows and perring (Homoptera: Aleyrodidae) in cotton. Southwest Entomologist 20:137–150

George DR, Callaghan K, Guy JH, Sparagano OAE (2008a) Lack of prolonged activity of lavender essential oils as acaricides against the poultry red mite (*Dermanyssus gallinae*) under laboratory conditions. Res Vet Sci 85:540–542. doi:10.1016/j.rvsc.2008.02.001

George DR, Smith TJ, Sparagano OAE, Guy JH (2008b) The influence of 'time since last blood meal' on the toxicity of essential oils to the poultry red mite (*Dermanyssus gallinae*). Vet Parasitol 155:333–335. doi:10.1016/j.vetpar.2008.05.005

Guimaraes JH, Tucci EC (1992) Evaluation of mineral oil in the control of *Dermanyssus gallinae* (De Geer, 1778) (Acari, Dermanyssidae) in field and laboratory conditions. Rev Bras Entomol 36:859–862

Höglund J, Nordenfors H, Uggla A (1995) Prevalence of the poultry red mite, *Dermanyssus gallinae*, in different types of production systems for egg layers in Sweden. Poult Sci 74:1793–1798

Hoop R (2008) Ektoparasiten. In: Aviforum (ed) Merkblätter Geflügelhaltung Aviforum, Zollikofen, Sect. 7.2. p 1–6

Isman MB (2006) Botanical insecticides, deterrents and repellents in modern agriculture and an increasingly regulated world. Annu Rev Entomol 51:45–66. doi:10.1146/annurev.ento.51.110104.151146

Kilpinen O, Roepstorff A, Permin A, Norgaard-Nielsen G, Lawsson LG, Simonsen HB (2005) Influence of *Dermanyssus gallinae* and *Ascaridia galli* infections on behaviour and health of laying hens (*Gallus gallus domesticus*). Br Poult Sci 46:26–34. doi:10.1080/00071660400023839

Kim S-I, Yi J-H, J-h Tak, Ahn Y-J (2004) Acaricidal activity of plant essential oils against *Dermanyssus gallinae* (Acari: Dermanyssidae). Vet Parasitol 120:297–304. doi:10.1016/j.vetpar.2003.12.016

Kim S-I, Na Y-E, Yi J-H, Kim B-S, Ahn Y-J (2007) Contact and fumigant toxicity of oriental medicinal plant extracts against *Dermanyssus gallinae* (Acari: Dermanyssidae). Vet Parasitol 145:377–382. doi:10.1016/j.vetpar.2006.12.021

Kirkwood AC (1967) Anaemia in poultry infested with red mite *Dermanyssus gallinae*. Vet Rec 80:514–516

Lamina J, Kruner N (1966) Die insektizide Wirkung hochdisperser Kieselsäuren auf Ektoparasiten des Geflügels. Dtsch Tierarztl Wochenschr 71:122–129

Lu Y, Fang J (2003) Bonferroni adjustment. In: Advanced statistics. River Edge, p 870

Maurer V, Perler E 2006. Control of the poultry red mite *Dermanyssus gallinae*. In: Organic congress 2006. Odense, May 30–31. 2006, pp 504–505

Maurer V, Baumgärtner J, Bieri M, Fölsch DW (1993) The occurrence of the chicken mite *Dermanyssus gallinae* (Acari: Dermanyssidae) in Swiss poultry houses. Mitt Schweiz Entomol Ges 66:87–97

Maurer V, Baumgärtner J, Fölsch DW (1995) The impact of *Dermanyssus gallinae* infestations on performance and on physiological and behavioural parameters of laying hens. In: Kropczynska D, Boczek J, Tomczyk A (eds) The acari. Physiological and ecological aspects of acari host relationships. European Association of Acarologists, Warszawa, pp 551–556

Momen FM, Amer SAA, Saber SA (2006) Acaricidal potentials of some essential, mineral and plant oils against the predacious mite *Neoseiulus cucumeris* (Oudemans) (Acari: Phytoseiidae). Acta Phytopathol Entomol Hung 41:383–393. doi:10.1556/APhyt.41.2006.3-4.20

Nordenfors H, Höglund J, Tauson R, Chirico J (2001) Effect of permethrin impregnated plastic strips on *Dermanyssus gallinae* in loose-housing systems for laying hens. Vet Parasitol 102:121–131. doi:10.1016/S0304-4017(01)00528-3

Palyvos NE, Athanassiou CG, Kavallieratos NG (2006) Acaricidal effect of a diatomaceous earth formulation against *Tyrophagus putrescentiae* (Astigmata: Acaridae) and its predator *Cheyletus malaccensis* (Prostigmatae: Cheyletidae) in four grain commodities. J Econ Entomol 99:229–236. doi:10.1603/0022-0493(2006)099[0229:AEOADE]2.0.CO;2

Rice ND, Winston ML, Whittington R, Higo HA (2002) Comparison of release mechanisms for botanical oils to control *Varroa destructor* (Acari: Varroidae) and *Acarapis woodi* (Acari: Tarsonemidae) in colonies of honey bees (Hymenoptera: Apidae). J Econ Entomol 95:221–226

Rosen S, Yeruham I, Braverman Y (2002) Dermatitis in humans associated with the mites *Pyemotes tritici, Dermanyssus gallinae, Ornithonyssus bacoti* and *Androlaelaps casalis* in Israel. Med Vet Entomol 16:442–444. doi:10.1046/j.1365-2915.2002.00386.x

Saez A, Fuentes Mora VH (2007) Comparison of the desiccation effects of marine and freshwater diatomaceous earths on insects. Stored Prod Res 43:404–409. doi:10.1016/j.jspr.2006.11.002

Thind BB, Ford HL (2006) Assessment of susceptibility of the poultry red mite *Dermanyssus gallinae* (Acari: Dermanyssidae) to some acaricides using an adapted filter paper based bioassay. Vet Parasitol 144:344–348. doi:10.1016/j.vetpar.2006.10.002

Zeman P, Zelezny J (1985) The susceptibility of the poultry red mite, *Dermanyssus gallinae* (De Geer, 1778), to some acaricides under laboratory conditions. Exp Appl Acarol 1:17–22. doi:10.1007/BF01262196

Variation in chemical composition and acaricidal activity against *Dermanyssus gallinae* of four eucalyptus essential oils

David R. George · Dino Masic · Olivier A. E. Sparagano · Jonathan H. Guy

Originally published in the journal Experimental and Applied Acarology, Volume 48, Nos 1–2, 43–50.
DOI: 10.1007/s10493-008-9225-z

Abstract The results of this study suggest that certain eucalyptus essential oils may be of use as an alternative to synthetic acaricides in the management of the poultry red mite, *Dermanyssus gallinae*. At a level of 0.21 mg/cm^2, the essential oil from *Eucalyptus citriodora* achieved 85% mortality in *D. gallinae* over a 24 h exposure period in contact toxicity tests. A further two essential oils from different eucalyptus species, namely *E. globulus* and *E. radiata*, provided significantly ($P < 0.05$) lower mite mortality (11 and 19%, respectively). Notable differences were found between the eucalyptus essential oils regarding their chemical compositions. There appeared to be a trend whereby the essential oils that were composed of the fewer chemical components were the least lethal to *D. gallinae*. It may therefore be the case that the complexity of an essential oil's chemical make up plays an important role in dictating the toxicity of that oil to pests such as *D. gallinae*.

Keywords *Dermanyssus gallinae* · Poultry red mite · Plant-derived product · Botanical pesticide

Introduction

The poultry red mite, *Dermanyssus gallinae* (De Geer), is currently the most economically deleterious ectoparasite of laying hens in Europe (Chauve 1998). Infestations of *D. gallinae* can result in significant stress to hens with subsequent reductions in bird condition, growth rate, egg quality and egg production (Chauve 1998). In extreme cases, mite population levels may be so high as to cause anaemia, and even death of hens (Wojcik et al. 2000; Cosoroaba 2001). Furthermore, *D. gallinae* may serve as a vector for numerous poultry pathogens (Chirico et al. 2003; Kim et al. 2007).

D. R. George (✉) · D. Masic · O. A. E. Sparagano · J. H. Guy
School of Agriculture, Food and Rural Development, Newcastle University, Newcastle upon Tyne NE1 7RU, UK
e-mail: D.R.George@ncl.ac.uk

Control of *D. gallinae* has typically been achieved through synthetic acaricides. The continued use of these products, however, may be hampered by issues of mite resistance (Kim et al. 2004; Fiddes et al. 2005) and reduced product availability as a result of more stringent legislation regarding existing and proposed active ingredients (Fiddes et al. 2005). In general the use of conventional cage systems for housing hens may alleviate the problem of *D. gallinae* infestation (Axtell 1999; Arkle et al. 2006), but these will be prohibited within the EU in 2012. It is therefore becoming increasingly important to identify alternative approaches for the management of *D. gallinae* in poultry production systems.

One such approach focuses on studying the toxic effects of plant-derived products for managing *D. gallinae*. Kim et al. (2004) tested 56 plant essential oils for their acaricidal effect on *D. gallinae*, and reported that 12 of these oils gave 100% mite mortality in contact toxicity tests at 0.07 mg oil per cm^2. Whilst this acaricidal effect against *D. gallinae* may be encouraging, the use of essential oils and/or botanical pesticides *per se* may be hampered by their inconsistence in efficacy, which has been reported as a reason for the poor market penetration of botanical insecticides to date (Isman 2008).

Chemical composition may differ between oils from the same or taxonomically similar species (Cimanga et al. 2002) due to external factors. These differences will in turn effect the biocidal activity of oils (Miresmailli et al. 2006), making it difficult to recommend any essential oil for its pesticidal qualities simply by taxonomy alone. As a consequence of this, it is important to consider differences in the pesticidal potential of plant essential oils from similar species when attempting to identify products that may be of use in pest management. Furthermore, where differences are found between seemingly similar oils, attempts should be made to assess the possible reasons for this variation so that further recommendations can be made on how best to source these products to ensure consistent effects.

With this in mind, the aim of the current study was to ascertain the acaricidal potential of essential oils from four eucalyptus species to *D. gallinae*. In addition, the chemistry of these four oils was assessed by Gas Chromatography Mass Spectrometry (GCMS) in an effort to explain any possible variation between oils regarding their acaricidal effect.

Materials and methods

The four eucalyptus essential oils used in the present study (Table 1) were all obtained by steam distillation and sourced from New Directions, Southampton, UK.

Table 1 Taxonomy and origin of the four eucalyptus species essential oils tested as acaricides against *Dermanyssus gallinae*

Species	Origin	Plant part used
Eucalyptus globulus (Labill.)	China	Leaves and wood
Eucalyptus radiata (Sieber)	Australia	Leaves
Eucalyptus citriodora (Hook)	China	Leaves and wood
Eucalyptus staigeriana (Muel)	Australia	Leaves and wood

Acaricidal effect

Essential oils were used to impregnate filter papers (Whatman No. 2, 4.25 cm diameter) at a rate of 0.21 mg/cm^2 (in 50 μl of ethanol, where control filter papers received 50 μl of ethanol only). According to treatment, these filter papers were then added to Petri-dishes (4.8 cm diameter × 1.2 cm height). After 3 min storage time in a fume cupboard to allow for evaporation of the ethanol, ~23 adult female *D. gallinae*, previously collected from a free range poultry unit, were added to each dish using a pooter and sealed therein. Mites were used within six days of collection so that only mites that had fed relatively recently were used. The Petri-dishes were stored for a period of 24 h at 22°C with a 16:8 light: dark cycle after which mortality was assessed under magnification. A mite was considered dead if no movement was observed following repeated agitation with an entomological pin (i.e. a purpose-made fine-pointed needle fitted with a handle). Percentage mortality of *D. gallinae* (with Abbots Correction to account for control mortality) was calculated for each essential oil. A total of four replicates were undertaken for each essential oil.

Following arcsin square-root transformation, the data were subjected to an analysis of variance (ANOVA), with corresponding Tukey's tests, to test for differences in percentage *D. gallinae* mortality between essential oils.

Essential oil composition

GCMS analysis of the four different eucalyptus essential oils was performed on a Hewlett–Packard 6890 GC split injector (280°C) linked to a Hewlett–Packard 5973 MSD (electron voltage 70 eV, source temperature 230°C, quad temperature 150°C multiplier voltage 2,000 V, interface temperature 310°C). The acquisition was controlled by a HP Kayak xa pc chemstation computer, in a full scan mode (35–535 amu/s). The sample (1 μl) in dcm was injected by an HP7683 auto sampler in split mode. After the solvent peak had passed (5 min) the GC temperature programme and data acquisition commenced. Separation was performed on a fused silica capillary column (30 m × 0.25 mm i.d) coated with 0.25 μm dimethyl-polysiloxane (HP-5). The GC was temperature programmed from 40 to 300°C at 4°C/min and held at final temperature for 1 min with Helium as the carrier gas (flow 1 ml/min, initial pressure of 50 kPa, split at 30 ml/min).

Results

Acaricidal effect

There was a significant difference in acaricidal effect between the essential oils from the four eucalyptus species ($F_{3,12} = 6.91$, $P < 0.01$). This resulted from significantly higher mortality ($P < 0.05$) when mites were exposed to the essential oil of *E. citriodora*, compared to when exposed to essential oil from *E. globulus* or *E. radiata* (Fig. 1). There was also a trend for mite mortality to be higher after exposure to the essential oil from *E. staigeriana* than after exposure to essential oil from *E. globulus* or *E. radiata*, but this was not statistically significant (where $P = 0.055$ and 0.132, respectively).

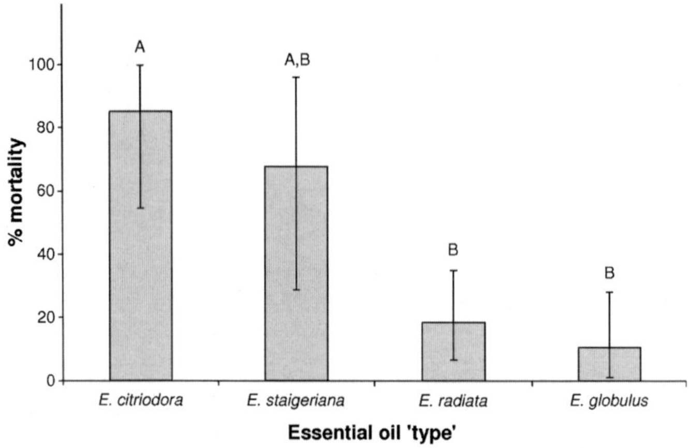

Fig. 1 Mean *Dermanyssus gallinae* mortality (with Abbots Correction) when exposed to different eucalyptus essential oils in 24 h contact toxicity tests at 0.21 mg/cm^2. Means and 95% confidence limits are back-transformed from data used in ANOVA $n = 4$ for all means. Pairs of means not accompanied by a common letter are significantly different according to Tukey's tests ($P < 0.05$)

Essential oil composition

As can be seen in Table 2, the chemical composition of the essential oils from *E. globulus* and *E. radiata* was similar. However, the composition of the oils from *E. citriodora* and *E. staigeriana* differed, both between these two oils and when compared to the oils from *E. globulus* and *E. radiata*.

When considering the total compounds constituting >0.3% of the essential oils, those from *E. citriodora* and *E. staigeriana* were more than twice as complex as the essential oil of *E. globulus*, and almost three times as complex as the oil from *E. radiata* (Table 2).

Discussion

Eucalyptus essential oil has been previously reported to possess biocidal properties and both *E. globulus* and *E. citriodora* essential oils have been tested for their acaricidal effect against the two-spotted spider mite, *Tetranychus urticae* (Koch.) (Choi et al. 2004). As in the present study, oil from *E. citriodora* proved to be more toxic to the mites than oil from *E. globulus*. Conversely, in a separate study that tested the toxic effect of essential oil from the same two species of eucalyptus against *D. gallinae*, oil from *E. globulus* was more toxic (100%) than that from *E. citriodora* (<50%) at 0.35 mg/cm^2 (Kim et al. 2004). Explanations for such varying toxicities of oils from a single or similar species are explored later, where factors such as growing conditions and harvesting/storage methods are considered.

In the present study, essential oils from *E. citriodora* and *E. staigeriana* proved to be relatively effective in killing *D. gallinae*, achieving more than 65% mite mortality. Conversely, essential oils from both *E. globulus* and *E. radiata* displayed a much reduced acaricidal effect, killing only 11 and 19% of mites after exposure, respectively. With the levels of variation in toxicity to mite species reported for eucalyptus essential oils elsewhere in the literature, it is perhaps not surprising that the acaricidal activity of the four oils

Table 2 Percentage contribution to the total oil of the main chemical components of four eucalyptus oils

Compound	*Eucalyptus globulus*	*Eucalyptus radiata*	*Eucalyptus citriodora*	*Eucalyptus staigeriana*
1,4-Cyclohexadiene	1.851	2.237	–	–
1,4-Methanoazulene	–	–	1.479	–
1,6-Octadien-3-ol	–	–	3.184	1.208
1,8 cineole	81.908	82.384	–	1.62
2,6-Octadien-1-ol	–	–	–	21.657
2,6-Octadienal	–	–	–	17.825
2,6-Octadienoic acid	–	–	–	5.304
3-Cyclohexen-1-methanol	–	1.848	–	
3-Cyclohexen-1-ol	–	–	–	2.527
6-Octen-1-ol	–	–	8.269	–
6-Octenal	–	–	55.336	–
alpha-Pinene	3.302	1.675	3.267	–
Benzene	3.374	2.615	–	1.613
Bicyclohex-2-ene	–	–	–	25.511
Caryophyllene			1.095	–
Cis-2, 6-dimethyl-2, 6-octadiene	–	–	2.908	–
Cyclohexanol	–	–	6.607	–
Cyclohexene	–	–	–	2.318
D-Limonene	7.606	6.692	1.435	–
Isopulegol	–	–	4.054	–
Total compounds (>0.3%) identified	10	8	23	23

Compounds contributing <1% of the total oil are not included. Similarly, compounds with <90% match are not included. Only compounds that were found to make up more than 0.3% of any complete essential oil are included within 'Total compounds (>0.3%) identified'

tested in the present study varied to this extent. The toxicity of other essential oils, even those derived from a single lavender species, has also been reported to vary elsewhere in the *D. gallinae* literature. The acaricidal effect of the oil from *Lavendula angustifolia* (Mill.) from Croatia was nearly 1.5 times more lethal to *D. gallinae* than that of essential oil from the same species, but from France (George et al. 2008).

GCMS analysis of the four eucalyptus essential oils revealed that *E. globulus* and *E. radiata* essential oils were similar, with the primary constituent in both being 1,8 cineole. The other two essential oils tested, however, showed no such trend for high 1,8 cineole content and appeared not only to differ in their chemical composition from *E. globulus* and *E. radiata*, but also from each other. The chemical composition of *E. globulus* and *E. citriodora* essential oils has also been evaluated elsewhere. Despite finding a reduced amount of 1,8 cineole in the former (44.3%) in comparison to the present study (81.9%), Cimanga et al. (2002) confirmed that this compound was absent from the oil of *E. citodora*. Whilst Cimanga et al. (2002) did not consider the essential oils from *E. radiata* and *E. staigeriana*, they did consider the chemical composition of a further eight essential oils from various *Eucalyptus* species, and as with the present study, found both similarities and differences between the oils tested. For example, *Eucalyptus tereticornis* (Smith) and *E. robusta* (Smith) shared similar quantities respectively of *β*-pinene (2.5 and 6.3%), *p*-cymene (28.6 and 27.3%) and cuminaldehyde (6.5 and 2.5%), whilst these components were found in notably different quantities in the oil of *E. alba* (Reinw.) (25.3, 7.4 and 0%, respectively). Variation in chemical composition has also been reported between essential oils obtained from different varieties or parts of a single plant (Moreno et al. 2007). In addition, numerous other factors may also effect the composition of essential oils. Geographic origin (Raal et al. 2007), seasonality (Flamini and Cioni 2007), method of oil extraction (Chiasson et al. 2001), year of harvest (Chalchat et al. 2007) and even storage conditions (Chalchat et al. 2007) are all factors that have been reported to influence essential oil chemistry.

With the chemical composition of essential oils being dependant upon so many factors, it is hardly surprising that results from different toxicity studies might not always agree. If the chemical composition of an oil is highly variable, then its toxic effect could be expected to be equally variable. Variation in chemical composition alone, however, may be an insufficient explanation for why some of the oils tested in the present study appeared to perform poorly as acaricides for *D. gallinae*. It might have been expected, for example, that the oils with the higher 1,8 cineole content would have performed well as acaricides, as this compound is commonly reported to have acaricidal (Macchioni et al. 2002; Miresmailli et al. 2006) or pesticidal (Yang et al. 2004) properties. It is possible that the increased toxicity of the essential oils from *E. citriodora* and *E. staigeriana* in the present study was attributable to their higher levels of octadien- and octen-based compounds, which were absent in detectable amounts in the oils from *E. globulus* and *E. radiata*, but there may also be an alternate explanation. The results suggest that the oils with the simpler chemical compositions were the least effective as acaricides for *D. gallinae*. It is possible that the components of the essential oils tested were operating in a synergistic manner to exert their toxic effect. In turn this could explain why the oils from *E. citriodora* and *E. staigeriana*, with their more complex chemical compositions, were more toxic to *D. gallinae* than the chemically simpler oils of *E. globulus* and *E. radiata* which contained fewer than half the total compounds (accounting for 0.3% or more of the oil) of the other two oils. Work elsewhere has shown that the acaricidal effect of complete essential oils is greater than combinations of their primary components, possibly because 'inactive' components act synergistically with 'active' components to achieve maximum toxicities (Miresmailli et al. 2006). This provides at least some support that essential oil chemical complexity may affect toxicity to pests, although this remains an area in need of further investigation.

Essential oils with more complex chemical compositions may have an additional advantage over simpler oils if developed for use as pesticides. It has been observed that one benefit of such products for pest management would be that the numerous active compounds in essential oils would make development of pest resistance to any essential oil-based product extremely difficult (Miresmailli et al. 2006). The aphid *Myzus persicae* (Sulz.), for example, was able to develop resistance to pure azadirachtin, a primary component of neem (*Azadirachta indica* A. Juss), but not to an extract of neem seeds containing a comparable quantity of this compound (Feng and Isman 1995).

In summary, whilst plant essential oils may hold promise in the management of pests such as *D. gallinae*, issues of variability in chemical composition are likely to complicate the use of these products. Indeed, it is unlikely that the poultry industry would accept any pest control product that could not be shown to be reliable in its effect. It may be the case that in order to produce a reliable, repeatable toxic effect from an essential oil, or product based upon an essential oil, it will not only be necessary to take account of the source plant's taxonomy to levels beyond that of even species, but also numerous other biotic and abiotic factors that could influence both the chemical composition of the source plant material and the oil produced from it. It may be possible to overcome some of these issues by standardising such variables as plant growing conditions, harvest times and storage protocols. For example, work suggests that techniques such as micro-propagation may be of use in reducing essential oil variability through encouraging the highly controlled proliferation of selected, reliable plant genotypes (Goodger et al. 2008), although this may prove to be inhibitory if it drives up the price of an essential oil and hence any final pest control product based upon it. Chemical standardisation and stricter quality control of various botanical insecticides might also be necessary if they are to compete with new generation synthetics (Isman 2008). Alternatively, the tendency for variation in essential oil chemical composition could be considered as advantageous. Growing conditions could be manipulated or sought to ensure high levels of chemical complexity within an oil, which according to the results presented might aid in maximising essential oil toxicity to pests such as *D. gallinae*.

Acknowledgments The authors would like to thank Mr. Paul Donohoe for GCMS analysis and Dr. Ed Okello for assistance with biochemical classification of essential oil constituents. This work was conducted as a satellite study to a project called MITEeHEN, for which funding is gratefully received from the Department of Environment, Food and Rural Affairs (UK).

References

Arkle S, Guy JH, Sparagano O (2006) Immunological effects and productivity variation of red mite (Dermanyssus gallinae) on laying hens–implications for egg production and quality. Worlds Poult Sci J 62:249–257. doi:10.1079/WPS200594

Axtell RC (1999) Poultry integrated pest management; status and future. Integr Pest Manag Rev 4:53–73. doi:10.1023/A:1009637116897

Chalchat JC, Ozcan MM, Dagdelden A, Akgul A (2007) Variability of essential oil composition of *Echinophora tenuifolia* subsp sibthorpiana Tutin by harvest location and year and oil storage. Chem Nat Compd 43:225–227. doi:10.1007/s10600-007-0087-0

Chauve C (1998) The poultry red mite *Dermanyssus gallinae* (De Geer, 1778): current situation and future prospects for control. Vet Parasitol 79:239–245. doi:10.1016/S0304-4017(98)00167-8

Chiasson H, Bélanger A, Bostanian N, Vincent C, Poliquin A (2001) Acaricidal properties of *Artemisia absinthium* and *Tanacetum vulgare* (Asteraceae) essential oils obtained by three methods of extraction. J Econ Entomol 94:167–171

Chirico J, Eriksson H, Fossum O, Jansson D (2003) The poultry red mite, *Dermanyssus gallinae*, a potential vector of *Erysipelothrix rhusiopathiae* causing erysipelas in hens. Med Vet Entomol 17:232–234. doi: 10.1046/j.1365-2915.2003.00428.x

Choi W, Lee S, Park H, Ahn Y (2004) Toxicity of plant essential oils to *Tetranychus urticae* (Acari: Tetranychidae) and *Phytoseiulus persimilis* (Acari: Phytoseiidae). J Econ Entomol 97:553–558

Cimanga K, Kambu K, Tona L, Apers S, De Bruyne T, Hermans N, Totté J, Pieters L, Vlietinck AJ (2002) Correlation between chemical composition and antibacterial activity of essential oils of some aromatic medicinal plants growing in the Democratic Republic of Congo. J Ethnopharmacol 79:213–220. doi: 10.1016/S0378-8741(01)00384-1

Cosoroaba I (2001) Massive *Dermanyssus gallinae* invasion in battery-husbandry raised fowls. Rev Med Vet (Toulouse) 152:89–96

Feng R, Isman MB (1995) Selection for resistance to azadirachtin in the green peach aphid, *Myzus persicae*. Experimentia 51:831–834. doi:10.1007/BF01922438

Fiddes MD, Le Gresley S, Parsons DG, Epe C, Coles GC, Stafford KA (2005) Prevalence of the poultry red mite (*Dermanyssus gallinae*) in England. Vet Rec 157:233–235

Flamini G, Cioni PL (2007) Seasonal variation of the chemical constituents of the essential oil of *Santolina etrusca* from Italy. Chem Biodivers 4:1008–1019. doi:10.1002/cbdv.200790069

George DR, Callaghan K, Guy JH, Sparagano OAE (2008) Investigating the residual toxicity of lavender essential oils as acaricides against the poultry red mite (*Dermanyssus gallinae*). Res Vet Sci 85:440–442. doi:10.1016/j.rvsc.2008.02.001

Goodger JQD, Heskes AM, King DJ, Gleadow RM, Woodrow IE (2008) Micropropagation of *Eucalyptus polybractea* selected for key essential oil traits. Funct Plant Biol 35:247–251. doi:10.1071/FP07241

Isman MB (2008) Botanical insecticides: for richer, for poorer. Pest Manag Sci 64:8–11. doi:10.1002/ps.1470

Kim S, Yi J, Tak J, Ahn Y (2004) Acaricidal activity of plant essential oils against *Dermanyssus gallinae* (Acari: Dermanyssidae). Vet Parasitol 120:297–304. doi:10.1016/j.vetpar.2003.12.016

Kim S, Na Y, Yi J, Kim B, Ahn Y (2007) Contact and fumigant toxicity of oriental medicinal plant extracts against *Dermanyssus gallinae* (Acari: Dermanyssidae). Vet Parasitol 145:377–382. doi:10.1016/j.vetpar.2006.12.021

Macchioni F, Cioni PL, Flamini G, Morelli I, Perrucci S, Franceschi A, Macchioni G, Ceccarini L (2002) Acaricidal activity of pine essential oils and their main components against *Tyrophagus putrescentiae*, a stored food mite. J Agric Food Chem 50:4586–4588. doi:10.1021/jf020270w

Miresmailli S, Bradbury R, Isman MB (2006) Comparative toxicity of *Rosmarinus officinalis* L. essential oil and blends of its major constituents against *Tetranychus urticae* Koch (Acari: Tetranychidae) on two different host plants. Pest Manag Sci 62:366–371. doi:10.1002/ps.1157

Moreno PHR, Lima MEL, Sobral M, Cláudia M, Young M, Cordeiro I, Apel MA, Limberger RP, Henriques AT (2007) Essential oil composition of fruit colour varieties of *Eugenia brasiliensis* Lam. Scientia Agricola 64:428–432. doi:10.1590/S0103-90162007000400014

Raal A, Orav A, Arak E (2007) Composition of the essential oil of *Salvia officinalis* L. from various European countries. Nat Prod Res 21:406–411. doi:10.1080/14786410500528478

Wojcik AR, Greygon-Franckiewicz B, Zbikowska E, Wasielewski L (2000) Invasion of *Dermanyssus gallinae* (De Geer, 1778) in poultry farms in the Torun Region. Wiad Parazytol 46:511–515

Yang YC, Lee H-S, Clark JM, Ahn YJ (2004) Insecticidal activity of plant essential oils against *Pediculus humanus capitis* (Anoplura: Pediculidae). J Med Entomol 41:699–704

Inert dusts and their effects on the poultry red mite (*Dermanyssus gallinae*)

Ole Kilpinen · Tove Steenberg

Originally published in the journal Experimental and Applied Acarology, Volume 48, Nos 1–2, 51–62.
DOI: 10.1007/s10493-008-9232-0

Abstract The haematophagous poultry red mite (*Dermanyssus gallinae*) is the most important pest of egg laying hens in many parts of the world. Control has often relied on chemical pesticides, but inert dusts, which are thought to kill target hosts primarily by desiccation, have become one of the most commonly applied alternative control methods for poultry red mite in Europe. This development has occurred despite a lack of knowledge of the efficacy of the different types of inert dusts and how this is affected by environmental parameters, e.g. the high relative humidity found in poultry houses. In this laboratory study the efficacy of different commercial inert dust products against *D. gallinae* is compared. All tested compounds killed mites, but there was a clear ranking of efficacy (measured as weight loss after 24 h and as time until 50% mortality), particularly at 75% relative humidity (RH). At 85% RH the efficacy was significantly lower for all tested compounds ($P < 0.001$). Weight changes over time followed an exponential evaporation model until the mites started dying whereafter the rate of evaporation increased again and followed a slightly different exponential evaporation model. A tarsal test showed that 24 h exposure to surfaces treated with doses much lower than those recommended by the producers is sufficient to kill mites as fast as when they were dusted with massive doses. These data emphasise the need for thorough treatment of all surfaces in a poultry house in order to combat *D. gallinae*.

Keywords Desiccation · Diatomaceous earth · Alternative control · Ectoparasite

Introduction

Dermanyssus gallinae (Acari: Dermanyssidae), the poultry red mite, is a blood-feeding mite of birds with a worldwide occurrence and is without any doubt the most important pest of egg laying hens in Europe (Chauve 1998). An estimated annual loss of 11 million Euros

O. Kilpinen (✉) · T. Steenberg
Department of Integrated Pest Management, The Danish Pest Infestation Laboratory,
University of Aarhus, Skovbrynet 14, DK-2800 Kgs. Lyngby, Denmark
e-mail: Ole.kilpinen@agrsci.dk

in the Dutch poultry sector alone emphasises the economic importance of this pest (van Emous et al. 2005). Infestation of a layer house with poultry red mites may cause significant irritation and anaemia in the birds and in severe cases may lead to death (Kilpinen et al. 2005). There have been reports of heavy mite infestations leading to reductions in egg production (Kirkwood 1967), and the mite is also known to be a possible vector of a number of avian pathogens, such as Newcastle Disease virus and avian spirochaetes and also of zoonotic bacteria such as Salmonella (Axtell and Arends 1990; Valiente-Moro et al. 2007). In addition, egg quality may be downgraded due to bloodspots from crushed mites on the eggshells (van Emous et al. 2005). Finally, mite infestation in a poultry house may be of nuisance to farm workers as it may cause dermatitis and severe pruritus (Auger et al. 1979; Höglund et al. 1995).

Dermanyssus gallinae generally hides in cracks and crevices in the near surroundings of the birds during most of the day and only spends brief periods of time on the birds for blood-feeding (Wood 1917). Control of poultry red mite has mainly relied on chemical pesticides, which has led to the development of resistance (Zeman 1987; Beugnet et al. 1997). In most countries these acaricides are only approved for usage in empty poultry houses in order to avoid chemical residues in eggs and poultry meat. This has severely restricted the possibilities for mite control and there is an urgent need for new and/or alternative control methods.

Inert dusts have been used for pest control for centuries, and so may provide just such an alternative. These dusts comprise a range of different minerals, diatomaceous earth (DE) products and synthetic silica products (Ebeling 1971; Subramanyam and Roesli 2000). Storage pests in particular have been treated with this type of agent, but a whole range of arthropod species have been targeted (Mewis and Ulrichs 2001; Faulde et al. 2006). In contrast, the treatment against poultry red mite using inert dust is a relatively new event. As the air humidity levels in poultry production facilities are often high this may hamper the efficacy of the inert dusts. Inert dusts mainly act as dessicants by absorbing the lipids of the cuticle surface (Ebeling 1971) leading to death of the arthropod as a result of water loss, so high humidity in general is less conducive to the action of inert dusts (Subramanyam and Roesli 2000; Nielsen 1998).

Inert dusts are being marketed for control of *D. gallinae* in many countries in Europe, and are frequently relied upon as one of very few legal measures for mite control among poultry farmers. However, very little information is available on the efficacy of different products and on the effect of different levels of relative humidity on the efficacy of these products, although previous work has included application of inert dusts against poultry red mite (Melichar and Willowitzer 1967 cited in Ebeling 1971); Tarshis 1967; Kirkwood 1974; Chirico 2004; Maurer and Perler 2006). The aim of this study was to compare the efficacy of different commercial inert dust products representing pure DE, DE modified with synthetic amorphous silica, and pure synthetic amorphous silica products. Two inert dust products not marketed for pest control, but with a potential acaricidal effect, were also included in the study to investigate if they also acted as desiccants.

Materials and methods

Mites

All experiments were conducted using blood-fed adult female poultry red mites from a culture kept in the laboratory since 1997 fed on regular layer hens. The mites were allowed to feed the night before the experiments to obtain a homogenous group of mites with regard to their water balance.

Inert dusts

Several different types of commercial products containing inert dust were obtained from producers or retail companies. The inert dusts included natural DE products: Insecto (Natural Insecto Products, Costa Mesa, CA, USA), Diamol (O.W.A. Kemi, Skanderborg, Denmark), and SilicoSec (Miljøfluen, Gandrup, Denmark), DE's modified with synthetic amorphous silica: ProtectIt (Hedley Technologies, ON, Canada) and Fossil Shield 90.0 (FS 90.0, Bein, Eiterfeld, Germany), and pure synthetic amorphous silica products: RID (Rentokil, Glostrup, Denmark) and ID (confidential formulation under test). In some of the experiments kaolin and talc were included for comparison. Kaolin and talc are porous inert compounds widely used as carriers or diluents in dry formulations of chemical and microbial pesticides and were anticipated to have a limited effect on the mites, although other similar types of minerals have been used with success against stored product pests (Subramanyam and Roesli 2000).

Evaporation experiments

Groups of 100 mites were collected in Pasteur pipettes and exposed to a surplus (sufficient product to completely cover all mites in the pipette with further product to spare) of the different types of inert dust. After thorough mixing of mites and inert dust, the mites were poured out again and transferred to clean glass vials with a fine metal mesh lid. Groups of control mites were treated similarly (collected and transferred to glass vials). The initial weight of each group of treated mites was determined by weighing (Sartorius RC 210D balance, lower limit 10^{-5} g). All glass vials were placed in a closed box with a constant relative humidity in a temperature controlled room at 25°C. The humidity in the box was kept at 75 or 85% RH by means of saturated solutions of NaCl or KCl, respectively. At variable intervals (ranging from 2 h intervals in the beginning of the experiment to once a week in the end) the glass vials were weighed individually and the relative weight losses were calculated based on the initial weight of the mites. At the same time, the number of dead mites in each vial was estimated. This was done by visual inspection according to the scale shown in Table 1. For the data analysis the scaling was transformed back to the averages for each group.

All mites came from a laboratory culture where they were likely to have mated and produced eggs in the days following feeding. However, as each vial contained 100 adult female mites, each of which can produce 5–10 eggs, it was not possible to determine accurately the number of offspring produced. It was noted whether the egg production was clearly reduced compared to the untreated controls or completely prevented.

Table 1 Scale for estimation of *Dermanyssus gallinae* mortality and the average number of dead mites used in the data analysis

Scale	Dead mites	Average number of dead mites.
0	0–5	2
1	6–20	13
2	21–79	50
3	80–94	87
4	95–100	98

In each experiment three groups of 100 mites were used for the individual treatments and the experiment was repeated twice, i.e. a total of 900 mites per treatment. However, it was not practically possible to include all treatments in the same experiment (on the same day) and as control groups were included in all experiments the total number of control mites was higher than those treated with the products.

Tarsal exposure tests

Mites were exposed to different doses of inert dusts on treated filter paper discs. Discs were treated in Langs Bell, an apparatus for treating surfaces uniformly with dry substances in powder or dust form (Lang and Melte 1930). Initially, the Langs Bell was calibrated by blowing in different amounts of material and weighing the amount deposited on filter paper discs (6 cm diameter = 28.3 cm^2). The calibration test showed that there was an approximately linear relationship between the amount blown into the bell (x) and the amount deposited on the paper (y): $y(x) = 0.0398x - 0.000009$ ($R^2 = 0.93$). Whilst there may have been some variation using this method, it was nevertheless more precise as compared to other application methods such as sieving (Collins and Cook 2006) or using static electric plastic dishes (Fields et al. 2003). As some of the doses applied in the tests were too small to be weighed for each paper individually, the linearity of Langs Bell was used to extrapolate the calibration curve to these lower doses.

Filter paper discs (6 cm diameter) were treated with three different doses (and one untreated control) of three different types of inert dust: Diamol, SilicoSec and ID, representing two natural DE products and one synthetic silica product. These products also represented low, medium and high efficacies based on the results of the evaporation studies. The doses were 0.085, 0.34 and 0.85 g/m^2.The treated paper discs were placed individually on a larger (7 × 7 cm) piece of fly glue (paper) trap (Cattle Shed, Silvanderson, Knäred, Sweden) and a group of 25 fed adult female mites were transferred to the filter paper disc. The bioassays were placed in plastic boxes with saturated solutions of NaCl or KCl to provide air humidities of 75 or 85% RH, respectively. After 24 h the remaining mites on the treated paper were sucked into Pasteur pipettes and kept in plastic boxes with the same relative humidity as before and observed daily for mortality over ~2 weeks or until all mites were dead. For each combination of inert dust product, dose and humidity, two groups of 25 mites were tested and the experiment was repeated once, i.e. a total of 100 mites per treatment.

Data analysis

In the evaporation experiment data were obtained on both weight change and mortality estimates. The weight loss after 24 h was noted and the data for the whole experimental period were fitted to a model for the relative weight change: $W(t) = W_{w0}$ [1 − exp(−kt) where W_{w0} is the water fraction of the weight at the start, and $-k$ is the rate of evaporation. This model is derived from the models for exponential water loss in Benoit et al. (2007). For the treatments with the slowest mortality (control, kaolin, and talc) the model was fitted only to those data before the mites started to die (the average time until the first mortality scores of 1). The remaining data were fitted to a slightly modified model: $W(t) = a(b - \exp(-kt))$ that does not necessarily intercept at 0% weight change.

The mortality scores were transformed back to estimated average mortality following Table 1. For day zero the mortality was by definition zero. The highest score of 4, corresponding to 95–100% mortality was recorded twice towards the end of the observation

Fig. 1 Examples of the weight changes of groups of *Dermanyssus gallinae* over time measured for **a** the untreated control mites at 75% RH, **b** kaolin at 85% RH, and **c** SilicoSec at 85% RH. The *curves* show the fitted exponential evaporation model $W(t) = W_{w0}\,(1 - \exp(-kt))$ for the entire experimental period (*dotted line*) and for the period until the mites start dying (*full line*). The *dashed line* shows the modified exponential model $W(t) = a(b - \exp(-kt))$ for the period from when the mites start dying

period where after the mortality was set as 100%. The estimated mortalities were fitted to a probit model (SAS Institute 2000) giving an estimate of the time elapsed following treatment until 50% of the mites were dead (LT_{50}).

Data on LT_{50} values and weight changes after 24 h for each of the two levels of relative humidity were log $(x + 1)$ transformed and tested for a significant effect of the treatment with a general linear model (SAS Institute 2000) followed by a multiple comparison analysis of the differences between the individual treatments (Tukey–Kramer, 0.05 significance level, SAS Institute 2000). The general linear model was also used to access differences between the two levels of humidity for each treatment.

Data from the tarsal exposure tests were analysed by calculating the LT_{50} for each experiment with the probit procedure (SAS Institute 2000). In some cases, particularly with control treatments at high humidity, no mites died during the 2 week observation period. Thus a relevant LT_{50} value could not be calculated and a fixed value of 50 days was used based on the observations in the evaporation studies. There was also a lower limit to the measure of LT_{50} because the first observation was made after 24 h. If all mites were dead by then, the LT_{50} is ~0.6 day which is the lowest possible with this set-up. These data were also log $(x + 1)$ transformed and tested for significant differences between the three treatments at each dose and RH (Tukey–Kramer, 0.05 significance level, SAS Institute 2000).

Results

Evaporation experiment

For the slowest acting treatments (control, kaolin and talc) there were clear deviations from the exponential evaporation model when the data was fitted to the entire experimental period (*dotted line* in Fig. 1). These deviations gradually became smaller as the treatments killed the mites faster. For all the commercial inert dust products the deviations were so small that the exponential evaporation model provided a good approximation of the weight changes over the experimental period. For the slow-acting treatments the model fitted the data well until the mites began to die (*solid line* in Fig. 1). For the remaining data the modified model, that did not necessarily start at 0% weight loss, gave a good fit with the data (*dashed line* in Fig. 1).

For both weight loss after 24 h and LT_{50} values, data varied significantly for the different types of treatment at both levels of relative humidity (general linear model, $P < 0.0001$ for all combinations). Particularly at the lowest humidity there was a clear grouping of the products with regard to mite weight loss after 24 h (Fig. 2: treatments with similar letters are not significantly different). Kaolin and talc caused slightly, but significantly higher weight loss (11.9 and 12.1%, respectively) compared to the control group (9.5%). For three products: Insecto, RID and Diamol, weight losses of around 20% were recorded (18.2, 20.1 and 21.1%, respectively). Slightly higher values were found for ProtectIt (25.9%) and SilicoSec (27.9%). Finally, the two products resulting in the highest weight losses were FS 90.0 (37.3%) and ID (52.9%). The results for LT_{50} data followed almost the same groupings, with high weight losses resulting in low LT_{50} values and vice versa.

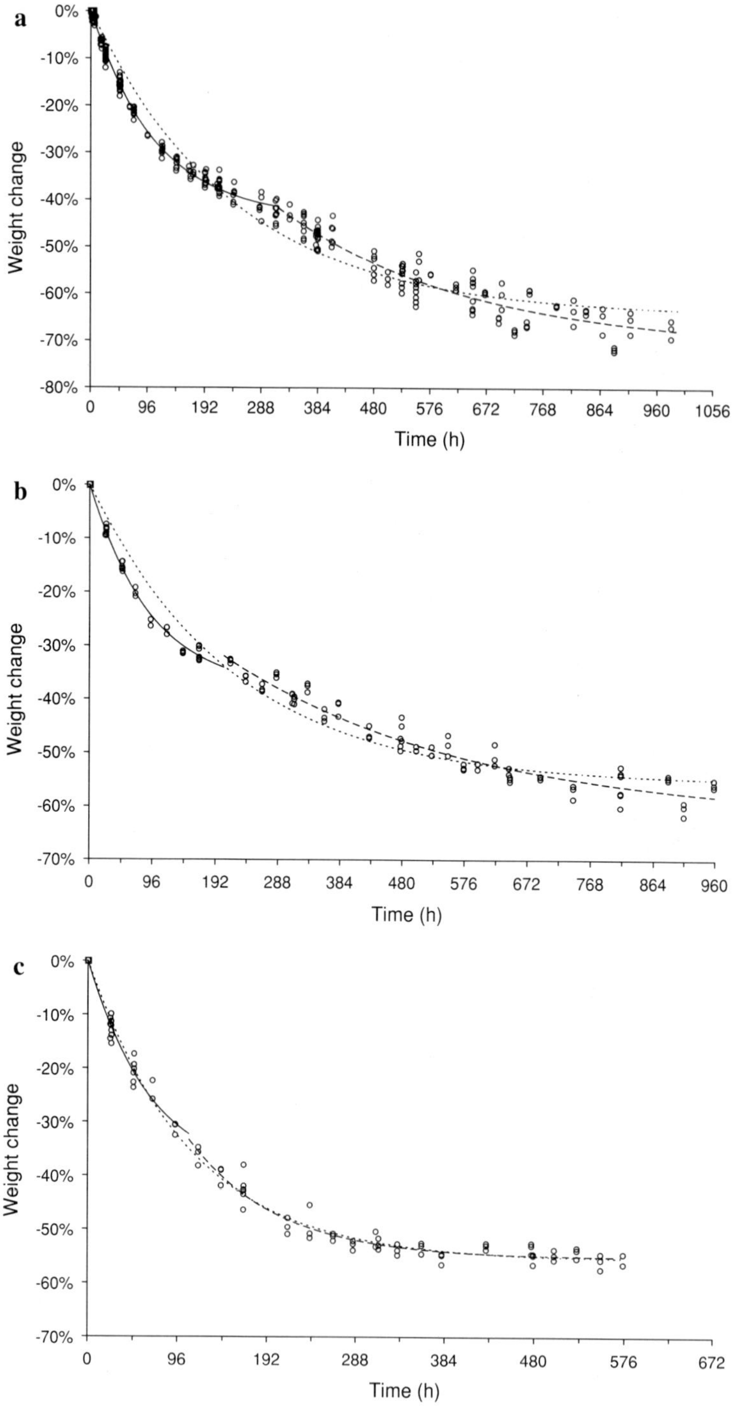
a
Weight change
0%
-10%
-20%
-30%
-40%
-50%
-60%
-70%
-80%
0 96 192 288 384 480 576 672 768 864 960 1056
Time (h)
b
Weight change
0%
-10%
-20%
-30%
-40%
-50%
-60%
-70%
0 96 192 288 384 480 576 672 768 864 960
Time (h)
c
Weight change
0%
-10%
-20%
-30%
-40%
-50%
-60%
-70%
0 96 192 288 384 480 576 672
Time (h)

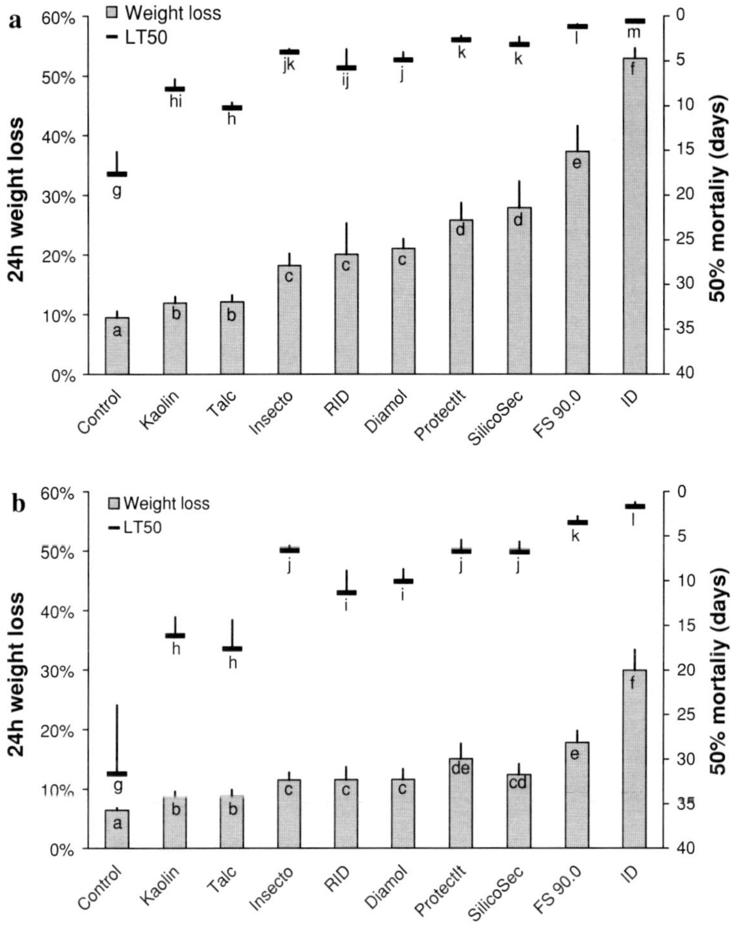

Fig. 2 Average weight loss of groups of 100 fed adult female *Dermanyssus gallinae*, relative to their initial weight before treatment and time until 50% mortality (for simplicity standard deviations are only shown in one direction). Results are provided from the two different levels of relative humidity used: **a** 75% and **b** 85%. *Bars* or *lines* with similar letters are not significantly different (0.05 level of significance, Tukey–Kramer multiple comparison, SAS Institute 2000). For all treatments there are significant differences for both weight change and LT_{50} at the two levels of humidity (GLM, SAS Institute 2000)

At 85% RH the weight losses were smaller and the LT_{50} values were higher, but the groupings remained almost the same as at 75% RH, although with more overlapping. One product, Insecto, showed a minor deviation from the general trend, in that the LT_{50} value was relatively lower than expected based on the weight change seen.

For all treatments there were significant effects of the humidity ($P < 0.001$ for all treatments) on both the LT_{50} values and weight change data; the LT_{50} being lower and the weight change being higher at 75% RH than at 85% RH.

At 85% RH, only the two most efficient products FS 90.0 and ID resulted in reduced egg production. At 75% RH, both of these products resulted in zero or very low egg production, and the five inert dusts (Insecto, RID, Diamol, ProtectIt, and SilicoSec) caused reduced egg production also. The kaolin and talc treatments had no observable effect on egg production.

In all experiments, the eggs that were produced hatched and the nymphs were not affected by the treatments.

Tarsal exposure test

There was a clear order of efficacy among the three inert dust products with ID being the most efficient and Diamol the least efficient (Fig. 3). However, with the lowest doses at 85% RH the differences between Diamol and SilicoSec are not significant. For all three products, increasing the dose or lowering the humidity results in lower LT_{50} values. It should be noted that the levelling out of the dose-effect at the highest doses of ID (particularly at 75% RH) was affected by the length of the observation interval (first observation after 24 h).

Comparing the LT_{50} values in this test with those obtained in the evaporation experiments (*arrows* in Fig. 3) showed that 24 h exposure on a treated surface (with the highest doses) was equally as efficient, if not more so, than the maximum exposure level achieved when mixing the mites with a surplus of dust.

Discussion

The potential of inert dusts for controlling insect pests, particularly stored product pests, has been known for centuries. The available literature on the effect of inert dusts on stored product pest covers a wide range topics, from efficacy (Mewis and Ulrichs 2001) to how the dust directly affects the insects (Ulrichs et al. 2006). To some extent this body of research covers the effect of these products on the Acari, particularly storage mites (Collins and Cook 2006) but little is known about how the inert dusts affect blood-sucking mites. Blood sucking ectoparasites such as the poultry red mite differ from the more commonly studied stored product pests because they consume relatively large amounts of avian blood which contains almost 90% water (Lehane 2005). An adult female poultry red mite ingests ~0.2 mg of blood per meal (Sikes and Chamberlain 1954). Despite the large amount of water ingested, the most efficient inert dusts in this study cause almost complete loss of water within 24 h at 75% RH (and 25°C) with the maximum dose applied in the evaporation studies. This indicates that the mites would not be able to compensate for this water loss by taking in a new blood-meal, at least not with the fastest acting products tested here. It is possible that with the more slowly acting products, additional blood-meals could replenish the mites' water loss, but further study would be needed to confirm this.

The initial weight loss of the untreated control mites can be attributed mostly to the defecation taking place in the first couple of days after feeding. The faeces remained in the glass tube but would have dried out with the water component passing through the mesh lid. This would have been the same for the treated mites but with additional water loss coming from evaporation across the mites' cuticle (Ebeling 1971). The present study has shown that the weight changes follow the exponential evaporation model of Benoit et al. (2007) with some modifications. For the untreated control mites the model fitted the weight changes until the mites started dying, then the rate of evaporation increased again indicating that the living mites had an active protection mechanism against evaporation. The weight changes from this point followed a new exponential evaporation model. With the faster acting treatments the transition point where the mites started dying was no longer detectable in the present experiment.

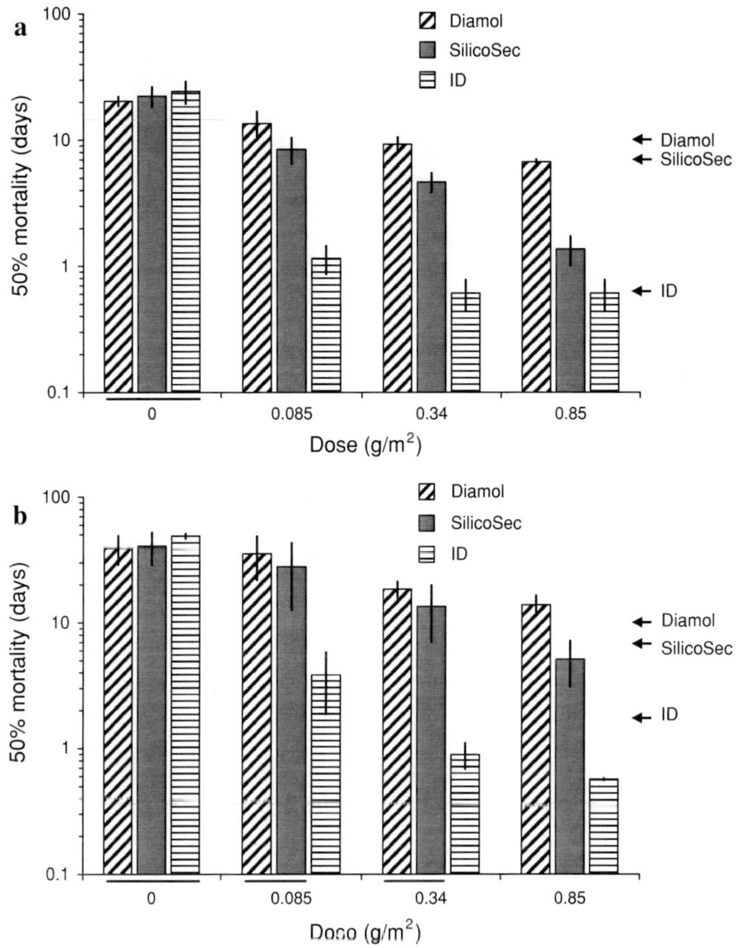

Fig. 3 Results from the tarsal exposure trials with *Dermanyssus gallinae*. Average LT_{50} values (with standard deviations) for the four tests at each dose (and untreated control) at 75% RH (**a**) and 85% RH (**b**). *Lines* below the *x*-axis indicate results that are not significantly different (0.05 level of significance, Tukey–Kramer multiple comparison on $\log(x + 1)$ transformed data, SAS Institute 2000). *Arrows* to the *right* indicate LT_{50} values from the evaporation studies. There is a lower limit to the LT_{50} determination at around 0.6 days due to the 24 h observation interval

There were clear and significant differences between the different types of inert dust, both with regard to weight loss and speed of mite knock-down. The most efficient product was the synthetic silica product ID that resulted in 50% mite mortality after 0.6 days at 75% RH and 1.7 days at 85%. These values may even be underestimating the efficacy of this product because knock-down was so quick that the observation intervals were too long to determine the LT_{50} precisely. The other types of inert dust fall in between the efficacy of ID and that of kaolin/talc as the least efficient compounds. The general rule concerning the efficacy of the different groups of inert dusts was as follows (least—most efficacious): clay and other minerals < pure DE's < modified DE's < pure synthetic amorphous silicas. This hierarchy is based mainly on the oil absorption capacity of these dust-based products which is dependent on their particle size distribution (Subramanyam and Roesli 2000). ProtectIt is a DE modified by the addition of a silica aerogel. However, in this experiment the pure DE, SilicoSec, was as efficacious as ProtectIt when comparing LT_{50} values, and SilicoSec killed mites significantly faster compared to another pure DE, Diamol. Similar results were found by Faulde et al. (2006) in a study on the German cockroach, *Blattella germanica*. FS 90.0 is also a modified DE, but has been modified by coating the DE particles by adding silica aerogel treated with dichlorodimethyl-silane. The results confirm that this produced a higher efficacy against poultry red mite. RID is an inert dust containing pure synthetic silicon dioxide like the product ID. However, while ID proved to be the fastest acting inert dust in the current study, RID performed relatively poorly. Kaolin and talc both proved to have a slight, but statistically significant acaricidal effect.

The general theory for the mode of action of inert dusts on arthropods is that they destroy the protective wax layer of the cuticle (Ebeling 1971), and it is possible that even such relatively harmless compounds as kaolin and talc absorbed some of the wax layer thus increasing the permeability of the cuticle. A study on the control of rice pests using entomopathogenic fungi showed that treatments in which the biocontrol agent was formulated in kaolin or talc provided significantly better as pest control products compared to untreated conidia (Samodra and Bin Ibrahim 2006). This might be an example of the inert dusts acting as synergists, as suggested by the authors. Perhaps more likely was that the compounds caused additional mortality to that caused by the fungus. According to Ebeling (1995), kaolin is worthless as a desiccant dust unless its pores are enlarged by acid- and heat-activation, and such derivates have been successfully applied against stored product pests (Permual and Le Patourel 1992). However, unmodified clays and other minerals have often been used as grain protectants against arthropod damage (Subramanyam and Roesli 2000). This suggests that dusts other than those containing high proportions of silicon dioxides may have pesticidal effects, although these are likely to be much reduced compared to synthetic silicon dioxides.

Exposing the mites for 24 h on treated surfaces confirmed that they were able to pick up enough product to be killed within less than a day for the most efficient product tested, ID, and 4.5 days (at 75% RH) with the less efficient Diamol. The doses applied in the current study were generally much lower than those recommended by companies marketing these inert dusts for use against poultry red mite or other pests. The recommended application rate of Fossil Shield is 1–3 g per layer hen in cage systems, which is notably higher than even the highest dose applied in the tarsal test assay. However, even at the relatively low doses used in this experiment, the efficacy of the products tested was comparable to that achieved in the evaporation experiment where massive doses of product where used in a maximum exposure procedure (Fig. 3). This suggests that the mites were susceptible to low doses of inert dusts provided they picked up the particles from a treated surface for a continuous period of 24 h. This emphasises the need for thorough treatment of all surfaces in a poultry house and also the need for reapplication where the treated surfaces become covered by environmental dust and debris.

Pest knock-down, or speed of kill, is an important factor when choosing between inert dust products, particularly if a product acts fast enough to prevent production of offspring. Another important factor is to what extent environmental variables such as humidity affects the efficacy of a product. A poultry house is a relatively humid environment and the humidity can rarely be controlled in layer houses other than by ventilation and temperature control. Nordenfors and Höglund (2000) recorded temperature fluctuations over 1 year in an aviary system. During summer the humidity fluctuated in the range 40–80%RH and in the winter the range was 60–85% RH. The results of the present study showed that the efficacy of all of the inert dusts tested was significantly lower at 85% RH compared to 75%

RH. In theory, synthetic silicon dioxide and modified DE's should be less dependent on lower humidity due to the hydrophobicity of these products. In the current study, however, all inert dusts showed decreased efficacy at the highest humidity level, so even the two tested pure synthetic silicas and the two modified DE's were susceptible to an increase from 75% RH to 85% RH. Mewis and Ulrichs (2001) observed a similar result for Fossil Shield in a study on stored product pests, and Faulde et al. (2006) found that SilicoSec, Diamol and Fossil Shield 90.0 were less effective than other modifications of Fossil Shield at 85% RH. Although wax sorption can take place in a water saturated atmosphere, provided the insect cuticle is in continuous contact with the dust (most notably for the silica aerogels), early reports also showed that knock-down will be prolonged compared to that achieved under ambient humidity conditions (Ebeling 1995).

In conclusion, based on the results of the current laboratory study, higher application rates of all the inert dusts, including the highly efficacious ones, must be applied if the humidity of the poultry house reaches 85% in order for these products to retain the efficacy shown at lower humidity levels. There are significant differences in the efficacy of the different products tested, suggesting that some may be more effective than others in poultry red mite control. Nevertheless, other factors including the price of any product and the ease of application would also need to be considered before selecting one product above another for application in poultry houses.

Acknowledgments We thank Minna Wernegreen, Nicolai Hansen, and Ib Bjarne Nielsen for technical assistance and Christos Athanassiou for supplying us with several inert dust products. This work is part of the SAFEHOUSE project financed by the European Commission contract no. FOOD-CT-2006-035547.

References

Auger A, Nantel J, Meunier N, Harrison JR (1979) Skin acariasis caused by *Dermanyssus gallinae* (de Geer): an in-hospital outbreak. Can Med Assoc J 17:700–703

Axtell RC, Arends JJ (1990) Ecology and management of arthropod pests of poultry. Ann Rev Ent 35:101–126. doi:10.1146/annurev.en.35.010190.000533

Benoit JB, Del Grosso NA, Yoder JA, Denlinger DL (2007) Resistance to dehydration between bouts of blood feeding in the bed bug, *Cimex lectlarius*, is enhanced by water conservation, aggregation, and quiescence. Am J Trop Med Hyg 76:987–993

Beugnet F, Chauve C, Gauthey M, Beert L (1997) Resistance of the poultry mite to pyrethroids in France. Vet Rec 140:577–579

Chauve C (1998) The poultry red mite *Dermanyssus gallinae*: current situation and future prospects. Vet Parasitol 8:364–376

Chirico J (2004) Kvalsterkontroll—spridningsvägar samt test av olika substanser och metoder. Fjäderfä 1:4 (in Swedish)

Collins DA, Cook DA (2006) Laboratory studies evaluating the efficacy of diatomaceous earths, on treated surfaces, against stored-product insect and mite pests. J Stored Prod Res 42:51–60. doi:10.1016/j.jspr.2004.09.002

Ebeling W (1971) Sorptive dusts for pest control. Ann Rev Ent 16:123–158. doi:10.1146/annurev.en.16.010171.001011

Ebeling W (1995) Inorganic insecticides and dusts. In: Rust MK, Owens JM, Reierson DA (eds) Understanding and controlling the German cockroach. Oxford University Press, Oxford, pp 193–226

van Emous RA, Fiks-van Niekerk TGCM, Mul MF (2005) Bloedluizen (vogelmijten) op papier en in de praktijk. PraktijkRapport 17 ISSN 15708624 (in Dutch, English summary)

Faulde MK, Tisch M, Scharningshausen JJ (2006) Efficacy of modified diatomaceous earth on different cockroach species (Orthoptera, Blattellidae) and silverfish (Thysanura, Lepismatidae). J Pest Sci 79:155–161. doi:10.1007/s10340-006-0127-8

Fields P, Allen S, Korunic Z, McLaughlin A, Stathers T (2003) Standardised testing for diatomaceous earth. In: Credland PF, Armitage DM, Bell CH, Cogan PM, Highley E (eds) Advances in stored products protection. Proceedings of the eighth international working conference on stored product protection. CABI International, Wallingford, pp 779–784

Höglund J, Nordensfors H, Uggla A (1995) Prevalence of the poultry red mite, *Dermanyssus gallinae*, in different types of production systems for egg layers in Sweden. Poult Sci 74:1793–1798

Kilpinen O, Roepstorff A, Permin A, Nørgaard-Nielsen G, Lawson LG, Simonsen HB (2005) Influence of *Dermanyssus gallinae* and *Ascaridia galli* infections on behaviour and health of laying hens (*Gallus gallus domesticus*). Br Poult Sci 46:26–34. doi:10.1080/00071660400023839

Kirkwood AC (1967) Aneamia in poultry infested with the red mite *Dermanyssus gallinae*. Vet Rec 80:514–516

Kirkwood AC (1974) Sorptive dusts for the control of the poultry red mite *Dermanyssus gallinae*. Int Pest Contr 16:12–15

Lang M, Melte E (1930) Zur Prüfung staubförmiger Erdflohmittel. Nachrichtenblatt für den Deutschen Pflanzenschutzdienst 10:11–12

Lehane M (2005) The biology of blood-sucking in insects. University Press, Cambridge

Maurer V, Perler E (2006) Silicas for control of the poultry red mites *Dermanyssus gallinae*. Proceedings paper presented at Joint Organic Congress, Odense, Denmark, 30–31 May, 2006 (reprint available: http://orgprints.org/7274/)

Mewis I, Ulrichs C (2001) Action of amorphous diatomaceous earth against different stages of the stored product pests *Tribolium confusum, Tenebrio molitor, Sitophilus granarius* and *Plodia interpunctella*. J Stored Prod Res 37:153–164. doi:10.1016/S0022-474X(00)00016-3

Nielsen PS (1998) The effect of a diatomaceous earth formulation on the larvae of *Ephestia kuehniella* Zeller. J Stored Prod Res 34:113–121. doi:10.1016/S0022-474X(97)00060-X

Nordenfors H, Höglund J (2000) Long term dynamics of *Dermanyssus gallinae* in relation to mite control measures in aviary systems for layers. Br Poult Sci 41:533–540. doi:10.1080/713654991

Permual D, le Patourel G (1992) Small bin trials to determine the effectiveness of acid activated kaolin against four species of beetles infesting paddy under tropical storage conditions. J Stored Prod Res 28:193–199. doi:10.1016/0022-474X(92)90041-N

Samodra H, Ibrahim YB (2006) Effectiveness of selected entomopathogenic fungi packed rice grain at room temperature against *Corcyra cephalonica* Stainton. Asean J Sci Tech Devel 23:183–192

SAS Institute (2000) SAS Users' Guide, release 8.01. SAS Institute, Cary, NC

Sikes RK, Chamberlain RW (1954) Laboratory observations on three species of bird mites. J Parasitol 40:691–697. doi:10.2307/3273713

Subramanyam BH, Roesli R (2000) Inert dusts. In: Subramanyam BH, Hagstrum DW (eds) Alternatives to pesticides in stored-product IPM. Kluwer, Doordrecht, pp 321–380

Tarshis B (1967) Silica aerogel insecticides for the prevention and control of arthropods of medical and veterinary importance. Angew Parasitol 8:210–237

Ulrichs C, Entenmann S, Goswami A, Mewis I (2006) Abrasive und hydrophil/lipophile Effekte unterschiedlicher inerter Stäube im Einsatz gegen Schadinsekten am Beispiel des Kornkäfers *Sitophilus granarius* L. Gesund Pflanz 58:173–181. doi:10.1007/s10343-006-0124-0

Valiente-Moro C, Desloire S, Chauve C, Zenner L (2007) Detection of *Salmonella* sp. in *Dermanyssus gallinae* using an FTA filter-based PCR. Med Vet Entomol 21:148–152. doi:10.1111/j.1365-2915.2007.00684.x

Wood HP (1917) The chicken mite: Its life history and habits. Bull 553 US Dept Agricult, Washington, DC

Zeman P (1987) Encounter the poultry mite resistance to acaricides in Czechoslovak poultry-farming. Fol Parasitol 34:369–373

Candidate predators for biological control of the poultry red mite *Dermanyssus gallinae*

Izabela Lesna · Peter Wolfs · Farid Faraji · Lise Roy · Jan Komdeur · Maurice W. Sabelis

Originally published in the journal Experimental and Applied Acarology, Volume 48, Nos 1–2, 63–80.
DOI: 10.1007/s10493-009-9239-1

Abstract The poultry red mite, *Dermanyssus gallinae*, is currently a significant pest in the poultry industry in Europe. Biological control by the introduction of predatory mites is one of the various options for controlling poultry red mites. Here, we present the first results of an attempt to identify potential predators by surveying the mite fauna of European starling (*Sturnus vulgaris*) nests, by assessing their ability to feed on poultry red mites and by testing for their inability to extract blood from bird hosts, i.e., newly hatched, young starlings and chickens. Two genuine predators of poultry red mites are identified: *Hypoaspis aculeifer* and *Androlaelaps casalis*. A review of the literature shows that some authors suspected the latter species to parasitize on the blood of birds and mammals, but they did not provide experimental evidence for these feeding habits and/or overlooked published evidence showing the reverse. We advocate careful analysis of the trophic structure of arthropods inhabiting bird nests as a basis for identifying candidate predators for control of poultry red mites.

Keywords Biological control · Ectoparasite · Poultry red mite · *Dermanyssus gallinae* · Predatory mites · *Androlaelaps casalis* · *Hypoaspis aculeifer* · European starling · *Sturnus vulgaris* · Chicken · *Gallus gallus* · Trophic structure · Bird nest · Poultry house

I. Lesna · M. W. Sabelis
Section Population Biology, Institute for Biodiversity and Ecosystem Dynamics, University of Amsterdam, Kruislaan 320, 1098 SM Amsterdam, The Netherlands

I. Lesna (✉) · P. Wolfs · J. Komdeur
Animal Ecology Group, Centre for Ecology and Evolution, University of Groningen, P.O. Box 14, 9750 AA Haren, The Netherlands
e-mail: lesna@science.uva.nl; i.k.lesna@rug.nl

F. Faraji
MITOX Trial Management BV, P.O. box 92260, 1090 AG Amsterdam, The Netherlands

L. Roy
Laboratoire de Parasitologie et Maladies Parasitaires, Ecole Nationale Vétérinaire de Lyon, Marcy l'Etoile, France

Introduction

The poultry red mite, *Dermanyssus gallinae* (De Geer) (Acari: Dermanyssidae), is a blood-sucking ectoparasite in nests of birds and small mammals. It is of economic importance as a worldwide pest in the poultry industry (Axtell and Arends 1990). This holds especially for 'laying hen' houses because the period of egg laying and breeding of domesticated chickens (*Gallus gallus*) is long (Maurer et al. 1993; Emous et al. 2005) relative to the time required for poultry red mites to double their population size (5.9 days at 25°C; Maurer and Baumgärtner 1992, 1994). In the Netherlands, outbreaks occurred infrequently in the past (mainly during summer), but now these occur throughout the year and in virtually all 'laying hen' (but also 'broiler breeder', 'rearing hen' and 'parent stock') houses. This increased incidence may be due to modern farming systems (e.g., constant climate) and perhaps also due to pesticide resistance arising from intensive chemical control. In addition, more strict allowance regulations have limited the number of pesticides available for control. As eggs are produced for human consumption, pesticides have to meet strict standards for food safety: (1) no residuals and (2) rapid breakdown into non-harmful components. Selection of effective chemicals is further complicated by the demand that pesticides should not harm bird health or bird welfare. Exposure of chickens to chemical sprays is hard to avoid, even though poultry red mites do not stay on their host permanently. This is because poultry red mites spend most of their time in the vicinity of the chicken and then they hide in crevices and other narrow places in the farm structure, where they are hard to target by pesticide sprays. The currently admitted pesticides do not suffice to eradicate poultry red mites.

Damage due to poultry red mites involves anaemia, increase in death risks and food demands, reduced time available for resting, decrease in disease resistance and egg-laying, and reduced egg quality (downgraded eggs due to egg shell with blood spots arising from squashed poultry red mites) (Emous et al. 2005). Based on a recent inquiry among farmers, the total costs from damage and control measures to the whole egg industry in The Netherlands are estimated to be 11 million Euros per year (Emous et al. 2005). This underestimates real costs because poultry red mites may vector disease agents of poultry (Valiente Moro et al. 2005, 2007) and they are a source of allergens causing dermatological problems to farmers and veterinarians (Rosen et al. 2002; Beck 1999).

Because current control methods are not sufficiently effective (Emous et al. 2005), we aimed to develop new methods of poultry red mite control by the use of their natural enemies, in particular predatory mites. This approach was pioneered by Buffoni et al. (1995, 1997) and Maurer and Hertzberg (2001). They reported the spontaneous occurrence of the predatory mite *Cheyletus eruditus* (Schrank) in the litter of poultry houses in Switzerland (for similar findings in Egypt, Mexico and UK, see: Abo-Taka 1996; Quintero and Acevedo 1984; Brady 1970a, b) and observed this mite feeding on juvenile poultry red mites. Releases of this predatory mite in poultry houses turned out not to yield control of poultry red mites, however. We pursued another approach to develop biological control methods by identifying predators of poultry red mites in their natural habitat. We analysed the food web structure of bird nests that have natural infestations of *D. gallinae*. Inventories of arthropods inhabiting bird nests have revealed a community structure involving bird parasites, microbivores and predators that may feed on them (e.g., Philips and Dindal 1979; Philips et al. 1989; Putatunda et al. 1989; Gupta and Paul 1989; Burtt et al. 1991; Lundqvist 1995; Kristofík et al. 1996; Philips 2000). Feeding at more than one trophic level (omnivory) may also be possible, but has not yet been shown for mites inhabiting bird nests. Some of the species are found exclusively in nests (nidicolous species), others are opportunistic visitors (e.g., edaphic species) and yet others (plant-dwelling species) end up in nests via plant material birds use for nest construction, and perhaps also via the green material birds use for courtship, for chemical control of ectoparasites or as a drug to boost their immune response (Clark 1991; Brouwer and Komdeur 2004; Gwinner and Berger 2005; Veiga et al. 2006). For the arthropods inhabiting a nest to establish a system with more trophic levels will require time. Hence, elaborate micro-ecosystems of arthropods are more likely to occur in association with birds that re-use nesting sites, such as European starlings (*Sturnus vulgaris* Linnaeus 1758). Our approach was therefore (1) to identify the mites inhabiting poultry houses and nests of European starlings, (2) to identify potential predators based on the literature and based on their numerical associations with *D. gallinae*, (3) to assess their predatory activity with respect to the blood parasite *D. gallinae* and microbivores, such as astigmatic mites, and finally (4) to test whether these candidate predators can switch to blood feeding on starlings and chicken in absence of prey (in which case they are omnivores because they feed at more than one trophic level).

Materials and methods

Mite inventory of poultry houses

Surveys of the mite fauna inhabiting four poultry farms in the provinces Brabant and Limburg (The Netherlands) were made every 2 weeks in the period from October 2007 to July 2008. Per sampling date and poultry house 20 samples were taken by Nordenfors traps, i.e., traps made of 3 mm thick, 140 × 100 cm, corrugated cardboard (Nordenfors and Chirico 2001). These traps were placed in laying nests and on perches, poles and walls. In addition, a variable number of samples from poultry house litter and from conspicuous *D. gallinae* aggregations were taken. Samples were inspected first under a binocular microscope and any mite suspected to be different from *D. gallinae* was collected in vials with alcohol, mounted in modified Hoyer's medium (Faraji and Bakker 2008) on a microscope slide and then identified. If available, at least 10 voucher specimens per species were maintained for later inspection.

Estimates of abundance were classified as rare (1–5 individuals/sample), common and never abundant (5–100/sample), common and sometimes abundant (>100/sample).

Mite inventory of starling nests

A total of 106 nest boxes, at least 6 m apart and 2.5 m above ground, were at our disposal at Vosbergen estate (Eelde-Paterswolde, The Netherlands). Because starlings re-use nest boxes and the mite fauna may become more diverse with time, the occupation of the nest boxes by breeding starlings was recorded in 2006 and the presence of old nest material was assessed in March 2007. A total of 29 nest boxes were occupied by starlings in April 2007, 14 of which had also been occupied in 2006. A large proportion of old nest material (from 2006 or earlier) was collected from each of all 29 nest boxes to assess the arthropod fauna. Nest boxes were not cleaned, however, to allow a significant proportion of the mite fauna to hide in grooves, cracks and crevices in the nest box. Within a week after fledging of the young starlings in June 2007, nest material was collected from the nest boxes. After transferring samples from the old (March 2007) and new (June 2007) nest material to Berlese funnels, the mites were collected in vials with alcohol over a period of 3–4 days, mounted in modified Hoyer's medium (Faraji and Bakker 2008) on microscope slides and then identified. If available, at least ten voucher specimens per species were maintained for later inspection.

Species-specific densities per nest were estimated and subsequently classified in the following ranges: 0–1; 2–10; 11–100; 101–1,000; 1,001–10,000; 10,001–100,000 individuals per nest. For statistical analyses these ranges were arbitrarily transformed into single discrete numbers, representing categorical densities: 1; 5; 50; 500; 5,000; 50,000.

Statistical analysis of mite–mite associations

Categorical densities were incorporated in statistical analyses as $\log_{10}$ transformed (categorical) variables. The statistical analysis was part of a more comprehensive multilevel modelling approach (MLwiN 2.10 beta 5 package; Rabash et al. 2008), with three hierarchical levels involving nest-box-group, nest box and nestling. A stepwise backward elimination was performed manually by removing the least significant independent fixed variable from the model.

As part of this modelling exercise, relationships of 'putative predator' densities (individuals per nest) were tested using a model with the variables occupancy history of nest box (occupied in 2006 or not) and *D. gallinae* density (individuals per nest) as well as three variables describing the state of the starlings occupying the nest box (brood size, hatching date, female quality; for definitions see Brouwer and Komdeur 2004). Since the emphasis in this article is on the inter-relationships between nest-inhabiting mites, we refrain here from describing the starling-related variables and discussing the results ensuing. This will be published elsewhere (P. Wolfs et al., in prep.).

Predation tests

Based on the nest inventory and the analysis of mite–mite associations, predatory activity was assessed for adult females of *Androlaelaps casalis* (Berlese). For reasons of comparison, we chose females of *Hypoaspis aculeifer* (Canestrini) because (1) it was occasionally found in starling nests, (2) it was observed to feed on *D. gallinae* (I. Lesna, pers. obs.) and (3) it was reared in our laboratory since 1991 (Lesna et al. 1995), and therefore readily available. Both *A. casalis* (obtained from starling nests in 2007) and *H. aculeifer* (obtained from lily fields in 1991) were cultured on a diet of *Tyrophagus putrescentiae* (Schrank), which in turn was reared on dry yeast flakes (22°C, 70% RH). One week before the predation test the cultures of both species of predatory mites were provided with ample prey to satiate the predators. Females nearing oviposition (i.e., showing a full grown terminal oocyte in their opisthosoma) were taken from the culture and transferred to vials (3 cm diameter, 4 cm high) with a moistened bottom of plaster of Paris mixed with charcoal. The vials had ten individuals of *D. gallinae* (mobile juveniles and adults). After transfer, the vials were tightly closed by a lid with an opening sealed with mite-proof gauze and placed in a climate room at 22°C, 70% RH and total darkness. After 24 h, numbers of live and dead *D. gallinae* were counted and all individuals were replaced by fresh ones to achieve the same starting density for a second day of the predation test. The predation experiments were replicated simultaneously in ten vials for each species and each day. Control experiments with vials containing *D. gallinae* alone were also carried out. Student *t*-tests for comparison of means were applied to detect differences in predation between the two species and between the 2 days of the experiment.

Apart from the predation assessment over two consecutive days we also made direct observations of more than 100 predation events and together with Prof. Urs Wyss (University of Kiel, Germany) we recorded part of these by the aid of a video-equipped binocular microscope. Stills of the video-records are included in this article (Figs. 3, 4).

Haematophagy tests

To test whether mites shown to have a capacity to prey upon *D. gallinae* also have a capacity to extract blood directly from the host of *D. gallinae*, we carried out two experiments one with starling and another with chicken. Three-day-old starlings and ten-day-old chickens were used to offer a bird stage most vulnerable to haematophagy (due to their thin skin and hence more accessible blood vessels). Moreover, given their less dense feather cover it was easier to observe bite marks, as well as mites especially on the starling host.

The experiments with young starlings were carried out in a brood incubator at 32–34°C (total darkness), installed at the field station in Vosbergen estate. Due to the proximity of the field station to the starling-occupied nest boxes, we were able to minimize the time between brood removal and their introduction into the incubator, as well as their reintroduction into their original nest. The young starlings were away from their nest no longer than 2.5 h. The experiments were carried out for 2 h during mid-day (13:00–15:00) to ensure that the young starlings had been fed by their parents before the experiment and that they would be fed by their parents before night. The young starlings were introduced each into a separate container (300 ml, 6 cm diameter) closed by a lid with a hole sealed with mite-proof gauze. Pieces of filter paper were provided on the bottom of the vial as a means to absorb moisture and to accommodate the young birds. Per container, 20 female mites of either *A. casalis*, *H. aculeifer* or *D. gallinae* were released that had been deprived of food for at least 3 days prior to the experiment. These three treatments were replicated five times and all replicates were carried out simultaneously. Bite marks on the young starling, gut colouration of the mites (as a bloodmeal indicator) and the presence of mites on and off the bird were checked just before and immediately after the 2 h experiment.

The experiments with young chickens were carried out from 18:00 to 12:00 next day in a climate room at 26 ± 2°C (darkness from 19:00 to 10:00 next day), at the experimental Farm Laverdonk (Heeswijk, The Netherlands). The experiment was carried out mostly during the night because *D. gallinae* is most active at night when the young chickens are sleeping. The young chickens were brought in cages with food and water in the climate room, the day before the experiment. At the start of the experiment the chickens were introduced each into a separate container (3 l volume, 17 cm diameter) closed by a lid with a hole sealed with mite-proof gauze and they had no access to food and water during the time spent in the container. Pieces of soft cardboard (from egg baskets) were provided on the bottom of the container as a means to absorb moisture and to accommodate the young birds. Per container 20 mites, nymphs and females of either *A. casalis*, *H. aculeifer* or *D. gallinae*, were released that had been deprived of food for at least 3 days prior to the experiment. These three treatments were replicated ten times. Bite marks on the young chickens, gut colouration of the mites and the presence of mites on and off the chickens were checked before and after the 18 h experiment.

Results

Mite inventory of poultry houses and starling nests

During summer time, the densities of the poultry red mite, *D. gallinae*, increased dramatically (up to 60.000 mobile stages per Nordenfors trap) with the time the laying hen flocks stayed in the poultry house (I. Lesna, pers. obs.). In individual starling nests, densities of *D. gallinae* could increase from a few individuals appearing at the time young starlings hatch to more than 30.000 at the end of the starling's breeding period (I. Lesna, pers. obs.). As argued in the Appendix, populations of *D. gallinae* from starling nests and poultry houses are conspecific. Blood-feeding mites other than *D. gallinae* were not found in poultry houses and starling nests, except for a few individuals of *Ornithonyssus sylviarum* (Canestrini and Fanzago) in a single starling nest (Roy et al. 2009).

Astigmatic mites varied in abundance largely depending on the presence of certain plant-derived materials, such as straw (collected from farmland nearby) and seeds in starling nests as well as chicken-feed in poultry houses. The species composition in starling nests was more diverse than in poultry houses (Table 1). Whereas three species of glycyphagid and histiostomatid species prevailed in starling nests, it was almost exclusively, the acarid mite, *Tyrophagus putrescentiae* (Schrank), that was found in poultry houses. The astigmatic mites are all thought to be microbivores and may serve as the main or alternative prey for various predatory mites.

Table 1 List of Acari found in association with starling nests (estate Vosbergen, Eelde-Paterswolde, Groningen) and in litter of poultry houses (Brabant, Limburg, The Netherlands)

Order Family	Species	Starling Nest	Poultry House
Mesostigmata			
Ascidae	*Blattisocius keegani* Fox	–	***
	Proctolaelaps pygmaeus (Müller)	–	*
	Proctolaelaps sp.[a]	*	–
	Zerconopsis remiger (Kramer)	***	–
Dermanyssidae	*Dermanyssus gallinae* (De Geer)[b]	****	****
Digamasellidae	*Dendrolaelaps fallax* (Leitner)	*	–
Laelapidae	*Androlaelaps casalis* (Berlese)	***	**
	Hypoaspis aculeifer (Canestrini)	*	–
Macrochelidae	*Macrocheles ancyleus* Krauss	*	–
Macronyssidae	*Ornithonyssus sylviarum* (Canestrini and Fanzago)	*	–
Parasitidae	*Parasitellus fucorum* De Geer[c]	*	–
Polyaspididae	*Uroseius acuminatus* (Koch)	–	*
Prostigmata			
Cheyletidae	*Cheyletus eruditus* (Schrank)	**	**
Tydeidae	*Lorryia reticulata* (Oudemans)	*	–
Astigmata			
Acaridae	*Aleuroglyphus ovatus* (Troupeau)	*	–
	Tyrophagus putrescentiae (Schrank)	–	**
	Tyrophagus longior (Gervais)	*	–
	Sancassania sp.	–	*
Glycyphagidae	*Glycyphagus domesticus* (De Geer)	**	–
	Lepidoglyphus destructor (Schrank)	**	–
Histiostomatidae	*Myianoetus* sp.[d]	**	–
Pyroglyphidae	*Dermatophagoides evansi* Fain, Hughes and Johnston	–	*
Winterschmidtiidae	*Saproglyphus* sp.[d]	*	–

* Rare, ** common, yet never abundant, *** common and sometimes abundant, **** very abundant. Mounted voucher specimens are available on request to I. Lesna or F. Faraji

[a] Possibly new species, currently being described (F. Faraji)

[b] Conspecific specimens from starling nests and poultry houses (L. Roy)

[c] A single nymph positively identified

[d] Unidentified species

Prostigmatic mites were represented almost exclusively by *Cheyletus eruditus* in both starling nests and poultry houses (Table 1). This species is known to be a predator of different species of astigmatic mites and can probably feed on *D. gallinae* (Maurer and Hertzberg 2001). We also noted dark-red coloured individuals collected from poultry houses where *D. gallinae* was very abundant. In starling nests, *C. eruditus* tended to be somewhat more abundant when there were more astigmatic mites, but the available data did not allow statistical analyses to test this claim.

After nestlings had fledged in June 2007, starling nests harboured mesostigmatic mites (other than *D. gallinae*) that were represented most frequently by *Androlaelaps casalis* and *Zerconopsis remiger* (Kramer) (Table 1). The latter species was absent in nest boxes before the start of the breeding season (March 2007), but *A. casalis* was then found frequently and abundantly in old nest material in ca. 80% of the nest boxes that had been occupied by starlings in 2006 (as opposed to ca. 30% in nest boxes that had not been occupied in the 2006). In some of the poultry houses sampled, *A. casalis* was also found in considerable numbers, *Z. remiger* was absent and *Blattisocius keegani* Fox was relatively the most abundant species (Table 1). Each of these four species can feed on astigmatic mites, such as *Acarus siro* (Linnaeus) and *T. putrescentiae* (I. Lesna, pers. obs.). Their potential to interfere with *D. gallinae* is further explored below. Strikingly, the mesostigmatic genera harbouring generalist predators, such as *Hypoaspis* spp., *Macrocheles* spp. and *Parasitus* spp., were rarely found in starling nests as well as in poultry houses.

Mite–mite associations

Based on visual inspection of scatter diagrams of the data obtained directly after the starling nestlings had fledged, we detected the following tendencies: densities of *A. casalis* and *Z. remiger* were relatively high when *D. gallinae* density was low, whereas densities of *A. casalis* and *Z. remiger* were low, when *D. gallinae* density was high. Because—just before nest building (March 2007)—*A. casalis* was more likely to be found in nest boxes that had nest material from 2006 (whereas then *Z. remiger* was absent), we further explored the negative relation between *A. casalis* and *D. gallinae* densities by taking the history of nest-box occupancy into account. Post-fledging densities of *A. casalis* in starling nests in June 2007 were significantly higher in nest boxes that had been occupied, than in nests that had no starling nest in 2006 (Fig. 1; Wald test $P = 0.013$). Post-fledging densities of *D. gallinae* were not significantly different between nest boxes that had been occupied in 2006 and those that had not been occupied in 2006 (Fig. 1; Wald test $P = 0.333$). Most strikingly, postfledging densities of *A. casalis* and *D. gallinae* in June 2007 showed a significant negative relationship in nest boxes that had a starling nest in 2006 (Fig. 2; Wald test $P < 0.001$), whereas there was no significant relationship in nest boxes that had not been occupied in 2006 (Fig. 2; Wald test $P = 0.414$).

This negative relationship requires an interpretation. It points at some form of interference between *A. casalis* and *D. gallinae*. Hence, one would expect variation in numbers among nests to be more reduced under conditions where the two interacting mite species are more likely to co-occur from the start of the starling's breeding period (provided that mite immigration rates are low compared to mite growth rates in the nest). These conditions may apply to nest boxes occupied by starlings in the previous year for the following reasons: (1) in these nests *A. casalis* is more likely to be present at the start of the starling's breeding period, and (2) the growth rates of *A. casalis* and *D. gallinae* are similar and most likely high (relative to migration into the nests). We hypothesize that these conditions prompted the significant negative relationship in nest boxes occupied in 2006 as well as the absence of such a relationship in nest boxes unoccupied in 2006.

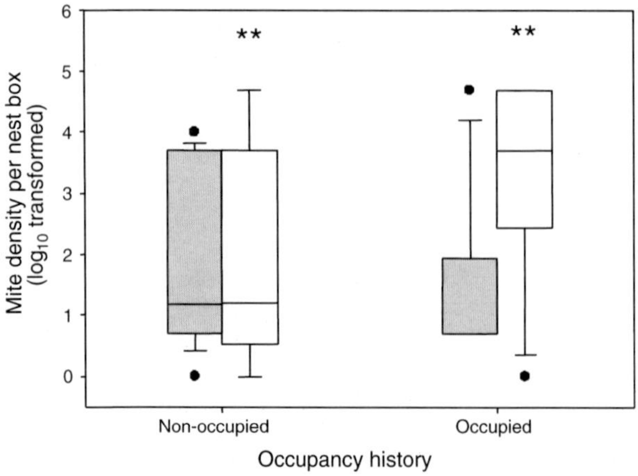

Fig. 1 Mite densities in nest boxes with different occupancy history. *Grey bars* represent *Dermanyssus gallinae* densities and *white bars* represent *Androlaelaps casalis* densities. Data for these densities were separated according to nest occupancy history; nest boxes which were not occupied ($n = 15$ for *D. gallinae*; $n = 14$ for *A. casalis*) and were occupied ($n = 14$ for both) by starlings in the previous breeding season (2006). The boxes in the figure represent the interquartile range, with the line within being the median. The error bars represent the 10–90th quartile range. Dots (●) represent data points interpreted as outliers. Note that the median (= 0.70) of *D. gallinae* in nest boxes occupied in the previous year coincides with the 10th quartile range. Significant differences between *A. casalis* densities from the N · O. category and those from the O. category are indicated by ** above the data ranges (Wald test; slope is equal to 1.235 with SE = 0.498 and is significantly different from zero at $P = 0.013$). Differences between *D. gallinae* densities from the two categories were not significant (Wald test; slope is equal to −0.519 with SE = 0.936 and is not significantly different from zero at $P = 0.333$)

The observed negative relationship between *A. casalis* and *D. gallinae* densities may arise from two distinct mechanisms: (1) competition for the same food resource (bird blood), (2) predation of one mite species on the other, or (3) a combination of competition and predation. To distinguish between these mechanisms, the most simple, first approach is to test whether *A. casalis* can feed on *D. gallinae* and whether it can extract blood from its bird host. The results of these predation and haematophagy tests are given below.

Predation tests

Predation tests were carried out with two species of mesostigmatic mites, one commonly found in starling nests and poultry houses, i.e., *A. casalis*, and one rarely found in these environments, i.e., *H. aculeifer*. Out of ten replicate experiments one female of *H. aculeifer* could not be retrieved after the second day (hence 9 replicates remained). For *A. casalis* two females could not be retrieved after the first day (hence 8 replicates remained). Control experiments with only *D. gallinae* did not show any *D. gallinae* mortality during the 2 days and are therefore left out of further analysis. The results of the experiments with putative predators of *D. gallinae* (Table 2) showed that the number of dead *D. gallinae* did not differ between day 1 and 2 of the experiment for *H. aculeifer* (*t*-test; $P = 0.35$) and for *A. casalis* (*t*-test; $P = 0.23$), but revealed significant differences between *H. aculeifer* and *A. casalis* treatments on day 1 ($P = 0.0012$) and day 2 ($P = 0.02$): the number of dead prey under exposure of *H. aculeifer* females was 1.8–2.1 times higher than that under exposure of *A. casalis* females. Oviposition was observed in several replicates, on both days for *H. aculeifer* and *A. casalis*. We refrained from quantifying oviposition because *A. casalis* and

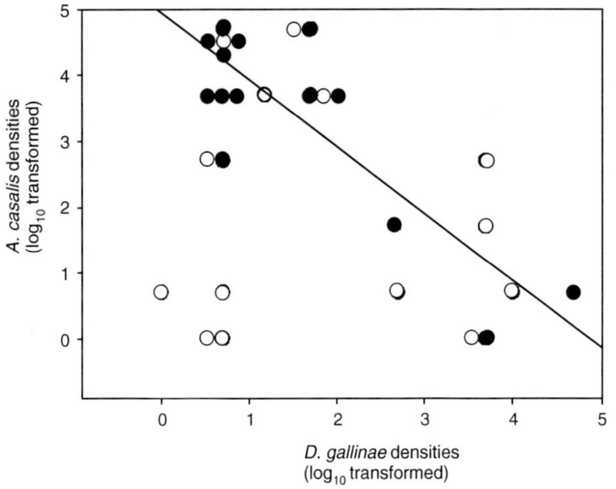

Fig. 2 Densities of *Androlaelaps casalis* plotted against *Dermanyssus gallinae*. Data for these densities were separated according to nest box occupancy history; nest boxes that were occupied (●)(n = 14) and were not occupied (○) (n = 14) in the previous breeding season (2006). The continuous line with negative slope represents the significant relationship between *A. casalis* densities ($\log_{10}$ transformed) and *D. gallinae* densities ($\log_{10}$ transformed) in nest boxes that were occupied in 2006 (Wald test; slope is equal to −1.016 with SE = 0.169 and is significantly different from zero at $P < 0.001$). For nests that were not occupied in 2006 the slope of the regression was not significantly different from zero (Wald test; slope is equal to −0.210 with SE = 0.257 and is not significantly different from zero at the 5% level since $P = 0.414$)

Table 2 Assessment of the rate of predation on poultry red mites (juveniles and adults) by females of two species of mesostigmatic mites during two consecutive days, following their rearing on a diet of astigmatic mites (*Tyrophagus putrescentiae*)

Species	Number of red mites killed per day							
	Day 1				Day 2			
	Mean	SD	Range	*n*	Mean	SD	Range	*n*
Hypoaspis aculeifer	5.5*ax*	1.7	3–8	10	6.4*ax*	2.5	3–10	9
Androlaelaps casalis	2.6*ay*	1.3	1–4	8	3.6*ay*	1.8	1–6	8

Climate room conditions were 22°C, 70% RH and total darkness; *n* = number of replicates; SD = Standard Deviation. Significant differences between means of two samples according to Student *t*-tests are indicated by different letters following the mean (between days: *a*, *b*; between species: *x*, *y*)

H. aculeifer have a strong tendency to hide eggs in small holes in the layer of plaster of Paris, and there were indications for egg retention in *A. casalis* females, a phenomenon known to occur in mesostigmatic mites under unfavourable conditions and reported for *A. casalis* by McKinley (1963).

Direct observations of attacks on *D. gallinae* individuals and ingestion of their body fluids were obtained for *H. aculeifer* (Fig. 3) and *A. casalis* (Fig. 4) during prey mortality assessments in the above tests and in more than 100 other predation tests. Attacks were observed on eggs and all mobile stages of *D. gallinae*. Large nymphs and adults were most frequently pierced at their flanks just behind the gnathosoma. Adults of *D. gallinae* were not sucked dry by the predatory mites. Instead they were only partially consumed. However, they were invariably leaking body fluids through the wound, were immobilized and

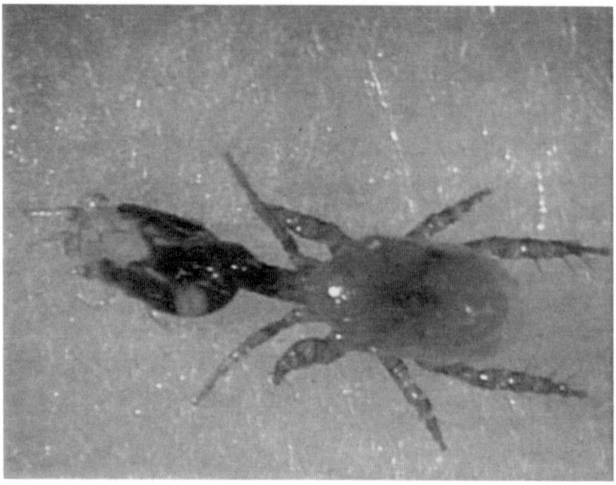

Fig. 3 Adult female of *Hypoaspis aculeifer* feeding on a nymph of the poultry red mite. Note that the gut of *H. aculeifer* is visible through the integument and starts to become *dark-coloured* (*red-brown*). Still from a video record made by Urs Wyss and Izabela Lesna. (Color figure online)

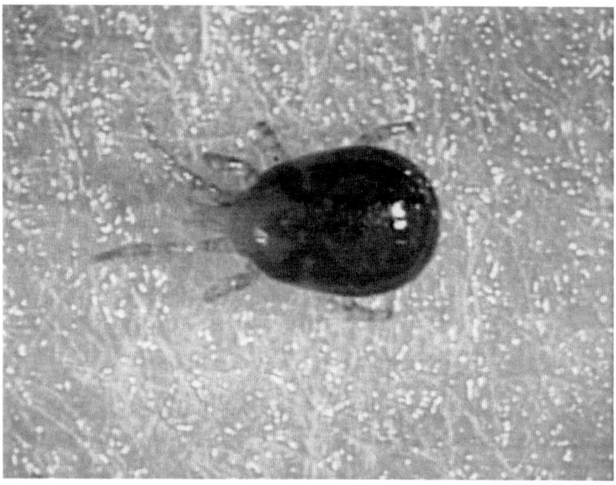

Fig. 4 Adult female of *Androlaelaps casalis* after feeding on a poultry red mite nymph, which causes their gut—visible through the integument—to turn red-brown (*dark-coloured* in this picture). Still from a video record made by Urs Wyss and Izabela Lesna. (Color figure online)

ultimately died. Due to the transparency of their integuments, body fluids were seen to move from the victim into the gut of *H. aculeifer* and *A. casalis* females, thereby causing the gut to assume a red-brownish colour (Figs. 3, 4). Such a change in gut colour does not occur when they fed on eggs or on unfed individuals of *D. gallinae*.

Haematophagy tests

Tests with starved *A. casalis* showed no evidence for haematophagy during 2 h of exposure to hatchlings of starlings at 32–34°C (Table 3). None of the mites were found on the host, none of them exhibited gut colouration (as in Fig. 4) and the host had no bite marks. In

Table 3 Replicated (n = 5) experiments to test whether starved females of two mite species feed on the blood of starling hatchlings (3 days after hatching)

Mite species	Replicate number	Number recovered	% with coloured gut
Dermanyssus gallinae	1	17	53
	2	12	58
	3	17	59
	4	19	68
	5	17	82
Androlaelaps casalis	1	17	0
	2	16	0
	3	17	0
	4	18	0
	5	15	0

Shown are the number of mites recovered after 2 h from the 20 individuals initially released per container and the percentage of these recovered mites with a red-brown coloured gut. Brood incubator conditions were 32–34°C, RH >50% and no daylight

contrast, tests involving *D. gallinae* yielded ca. 10% of the mites on host, 50–80% of the mites with red-brown coloured guts (indicating a fresh bloodmeal) and some hosts with bite marks (Table 3). Unfortunately, *H. aculeifer* did not survive the conditions of this experiment. Separate trials showed that temperatures above 30°C are detrimental to survival of *H. aculeifer*. For this reason the haematophagy tests with young chicken were carried out at a lower, yet bird-friendly temperature (26 ± 2°C). These tests (Table 4) showed that both *H. aculeifer* and *A. casalis* females cannot obtain blood during 18 h of exposure to young chicken. In contrast, 50–90% of the *D. gallinae* that were recovered had red-coloured guts. Possibly, because the light had been switched on 2 h before collecting the mites from the containers, none of the *D. gallinae* mites were found on the hosts. Bite marks on the young chicken were noted in only few cases, but they were not easy to observe because 10-day-old chicken possess already a more dense feather cover.

Not all of the 20 mites released per container were recovered at the end of the experiments. In the tests with starling hatchlings 12–19 *D. gallinae* and 15–19 *A. casalis* were recovered. The missing individuals are most likely present in or under the starling's droppings, where part of them may have gone unnoticed. Escape during inspection and hiding on the host is quite unlikely for these experiments. In the tests with young chickens relatively more mites (especially *D. gallinae* and *A. casalis*) were missing for a variety of reasons, the most likely of which was that they were hiding in the cardbox structure and chicken faeces on the bottom of the containers. Probably due to the relatively larger size of *H. aculeifer* retrieval of released mites was less of a problem (10–18 mites recovered, as opposed to 4–14 for the other 2 species).

Discussion

Our approach was to explore species of putative *D. gallinae* predators that occur in association with *D. gallinae* in bird (starling) nests under natural conditions, and those that occur spontaneously in poultry farms in which *D. gallinae* is a pest. Two species, one laelapid (Mesostigmata) and one cheyletid (Prostigmata), were found more or less frequently in large numbers in both environments: *Androlaelaps casalis* and *Cheyletus eruditus* (Table 1). Two other species, both ascids (Mesostigmata), were found in only one of the

Table 4 Replicated ($n = 10$) experiments to test whether starved females of three mite species feed on the blood of young chicken (10 days since hatching)

Mite species	Replicate number	Number recovered	% with coloured gut
Dermanyssus galinae	1	11	91
	2	12	50
	3	4	25
	4	7	86
	5	10	70
	6	11	73
	7	12	75
	8	10	80
	9	10	60
	10	8	62
Hypoaspis aculeifer	1	18	0
	2	17	0
	3	10	0
	4	15	0
	5	13	0
	6	14	0
	7	12	0
	8	11	0
	9	14	0
	10	18	0
Androlaelaps casalis	1	12	0
	2	14	0
	3	11	0
	4	6	0
	5	15	0
	6	8	0
	7	8	0
	8	14	0
	9	11	0
	10	6	0

Shown are the number of mites recovered after 18 h from the 20 individuals initially released per container and the percentage of these recovered mites with a red-brown coloured gut. Climate room conditions were 26°C, RH >50% and 15 h of darkness

two environments and then sometimes in large numbers: *Zerconopsis remiger* in starling nests and *Blattisocius keegani* in poultry farms (Table 1). Strikingly, another laelapid species, *Hypoaspis miles* (Berlese) (probably *Stratiolaelaps scimitus* (Womersley); F. Faraji, pers. obs. 2008), currently used to control *D. gallinae* and other blood feeding mites (e.g., the snake mite, *Ophionyssus natricis* (Gervais)) on pet animals, such as canaries, pigeons and reptiles (J. Evers, REFONA BV, pers. comm. 2008), was never found spontaneously in starling nests as well as in poultry farms. Only a few individuals of yet another laelapid species, *Hypoaspis aculeifer*, were found at the end of the breeding season in two starling nests at Vosbergen estate (Eelde-Paterswolde, The Netherlands). This raises the question whether the putative *D. gallinae* predators found in association with *D. gallinae* in bird nests under natural conditions offer perspectives for control of *D. gallinae* and how they compare to *H. miles*, currently used in practice for control of *D. gallinae* on pet animals but not (yet) in poultry farms. This question needs to be answered in future experiments in cages and poultry farms. Below, we discuss the arguments as to why the two *Hypoaspis* species and one of the four species of putative *D. gallinae* predators from starling nests, *A. casalis*, represent candidate predators for biocontrol of *D. gallinae* in poultry houses.

The laelapid mites, *H. miles* and *H. aculeifer*, are mainly ground-dwelling predators and they are occasionally reported to occur in nests of birds in low to very low numbers (e.g., Gwiazdowicz et al. 1999; Kristofík et al. 2003; Fenda and Lengyel 2007). Yet, they seem to be somewhat more numerous in nests of ground-nesting birds, such as European bee-eaters and sand martins (Kristofík et al. 1996). We found *H. aculeifer* in nest boxes with starling nests but then only very late in the breeding season and in very low numbers, even when *D. gallinae* infestations emerged early in the breeding season. All these observations do not lend support to the hypothesis that *H. miles* and *H. aculeifer* have a strong association with bird nests, let alone with *D. gallinae* infestations. The observations are more consistent with the hypothesis that they inhabit soil litter and visit bird nests. Testing whether *H. miles* and *H. aculeifer* in the soil litter exhibit an aggregative response to the density of *D. gallinae* in nests of ground-nesting birds, has not yet been done, but it seems a feasible and informative experiment to do.

Both *H. miles* and *H. aculeifer* may well be opportunistic generalist predators of *D. gallinae* under natural conditions. We found strong evidence for *H. aculeifer* females attacking and feeding on all stages of *D. gallinae* (Table 2; Fig. 3). In fact, starved females of *H. aculeifer* are voracious predators once released in Petri dishes with *D. gallinae* and they resume oviposition within a day (I. Lesna, pers. obs.). Also, *H. miles* has proven to be a voracious predator of *D. gallinae* and has proven to reproduce on an exclusive diet of *D. gallinae* (Tuovinen 2008; J. Evers, pers. comm. 2008). For reasons unclear to us, *H. miles* and *H. aculeifer* have never been considered to be omnivores that feed on *D. gallinae* as well as the blood from the host of *D. gallinae*. Clearly, the mouthparts of these species exhibit none of the adaptations known for true blood-feeding acarines. Our blood-feed tests with *H. aculeifer* and young chicken also did not provide any evidence for feeding on blood of the chicken, even when in a stage where it is most vulnerable to blood-feeding ectoparasites (Table 4). We therefore conclude that *H. aculeifer* and probably also *H. miles* are true predators and candidate agents for biocontrol of *D. gallinae* in poultry houses. The only cautionary remark is that *H. aculeifer* (in contrast to *A. casalis*) cannot survive at 32–34°C in the brood incubator where we carried our blood-feed tests on young hatchlings of European starlings and that such temperatures do occur on hot summer days in Dutch poultry houses (N. Harteveld, pers. comm. 2008).

The laelapid mite, *A. casalis*, has been reported as a frequent and sometimes abundant inhabitant of the nests from a wide variety of bird species: white storks (Bloszyk et al. 2005), white-tailed sea-eagles (Fenda and Lengyel 2007), owls (Kristofík et al. 2003; Gwiazdowicz 2003), eagles, harriers, buzzards, kites, ospreys, falcons (Gwiazdowicz 2003), house wrens (Pacejka and Thompson 1996; Pacejka et al. 1998), woodpeckers (Pung et al. 2000), European bee-eaters (Kristofík et al. 1996), reed warblers (Kristofík et al. 2001) and European starlings (this article). Hence, its common name is 'the cosmopolitan nest mite'. It has also been found in poultry houses in the UK, Egypt and The Netherlands in considerable numbers (McKinley 1963; Brady 1970a, b; El-Kammah and Oyoun 2007; this article).

Its trophic position in the food web of organisms in bird nests has been unclear. McKinley (1963) observed *A. casalis* feeding on droplets of human blood, but is of the opinion that *A. casalis* cannot penetrate mammalian or avian skin and can therefore not be a blood parasite, a view shared by Hughes (1976) and Tenquist and Charleston (2001). Indeed, the structure of its gnathosoma and in particular its chelicerae are very different from other blood-sucking mesostigmatic mites, such as Dermanyssidae (*Dermanyssus gallinae*; McKinley 1963; Roy and Chauve 2007), Macronyssidae (*Ornithonyssus sylviarum*; Evans 1957) and Rhinonyssidae (*Sternostoma tracheacolum* Lawrence; Evans 1957). Other authors assume that *A. casalis* is a (facultative) blood parasite (Men 1959; Radovsky 1985, 1994; Kristofík et al. 1996; Pacejka and Thompson 1996; Pacejka et al. 1996, 1998; Pung et al. 2000; Phillips 2000; Rosen et al. 2002; Svana et al. 2006; Fenda and Lengyel 2007; for review, see Proctor and Owens 2000). However, despite the presence of *A. casalis* in bird nests in relatively high numbers negative effects on brood performance have not been found in the house wren (Pacejka and Thompson 1996; Pacejka et al. 1998) and in the red-cockaded woodpecker (Pung et al. 2000). This may indicate that blood parasitism is not the predominant mode of feeding in *A. casalis*, or it may even be absent altogether. Some authors consider the possibility that *A. casalis* is a predator of small arthropods (including mites) in addition to being a blood parasite (Pacejka and Thompson 1996; Pacejka et al. 1996; Kristofík et al. 1996). Using a statistical method called path analysis, Pacejka et al. (1998) found a positive, direct effect of the numbers of *Dermanyssus hirundinis* (Hermann) on the numbers of *A. casalis*. Referring to Pacejka et al. (1998), Proctor and Owens (2000) hypothesize *A. casalis* to be a predator of blood-feeding mites and therefore a potential mutualist to the nesting bird. Pacejka et al. (1998) did not consider the effects of arthropods other than blood feeding mites and overlooked the possibility that *A. casalis* might be a predator of non-parasitic mites as well as young, vulnerable stages of parasitic and non-parasitic insects in bird nests. For example, *A. casalis* can feed on several astigmatic mite species (McKinley 1963) and has been considered as a biocontrol agent of astigmatic mites that are pests in stored products (Barker 1968). Various species of astigmatic mites in stored products are also found in bird nests, where they are probably feeding on fungi. Clearly, the trophic relations of *A. casalis* with other nest-dwelling arthropods and with the bird are in need of a causal experimental analysis.

In this article we provide quantitative evidence supporting the hypothesis that *A. casalis* is a predator of the poultry red mite (*D. gallinae*) (Table 2; Fig. 4) and of astigmatic mites living in nests (*Glycyphagus domesticus* (De Geer), *Lepidoglyphus destructor* (Schrank)) and in poultry houses (*Tyrophagus putrescentiae*). We also show quantitatively that *A. casalis* could not extract blood from young chicken and young starlings (Tables 3, 4), i.e., from birds in a life stage where they are most accessible and vulnerable to acarine blood parasites. These results are largely in agreement with the qualitative observations reported by McKinley (1963). This author also observed starved *A. casalis* feeding on astigmatic mites and poultry red mites and found no evidence (based on mite gut colouration or bite marks on host skin) for feeding on chicken, laboratory mice and men (even when the human skin was treated with fine sandpaper to improve access to blood vessels). Thus, given its numerical abundance in starling nests, its inverse association with *D. gallinae* in starling nests, its spontaneous occurrence in poultry houses infested with *D. gallinae,* its ability to complete its life cycle on a diet of *D. gallinae* and its inability to acquire blood directly from a bird (even when in its most vulnerable stage), we conclude that *A. casalis* is a true predator. This conclusion should challenge ornithologists to revise their views on how to judge (and perhaps how birds judge) the risk of ectoparasitism in a nest and their views on why birds re-use old nesting sites or old nest material (e.g., Mazgajski (2007) considers old nest material of European starlings only as a source of ectoparasites). Moreover, it points at a new candidate agent for the control of *D. gallinae* in poultry houses.

Apart from *H. miles*, *H. aculeifer* and *A. casalis*, there may well be several other candidate predators for biocontrol of *D. gallinae*. Our inventory of mites in starling nests and poultry farms in The Netherlands yielded species that may act as predators of *D. gallinae*, like *C. eruditus*, *Z. remiger* and *B. keegani*. Also, our attempt to review the literature on mites in bird nests appeared to provide several nidicolous species, such as *Hypoaspis lubrica* Voigts and Oudemans (Brady 1970a, b; Gwiazdowicz et al. 1999). However, none of these species have been actually tested for their ability to feed on *D. gallinae*, nor for their (in-)ability to feed on the blood of their hosts. Some of these species have been observed to feed on free blood (e.g., *H. lubrica* feeding on free blood of mice; Li and Meng 1992), but as shown for the case of *A. casalis* feeding on free blood is not sufficient evidence to infer haematophagy (McKinley 1963; this article).

As much as we realize that our conclusions on the exclusive predatory life style of *A. casalis* and *H. aculeifer* may not hold under all conditions (e.g., other bird species), we caution against uncritical citing and unconfirmed inferences on feeding life styles of nidicolous arthropods (e.g., Pacejka et al. 1998; Xing-Yuan et al. 2007). Bird nests may not only harbour ectoparasites and fungivores but also predators of one or even both of these trophic guilds. Indeed, the arthropods in nests are part of a multitrophic system.

Acknowledgments We (IL, MWS) thank René Heijnen (CeHaVe, Veghel, The Netherlands) and Karel Bolckmans (Koppert BV, Berkel en Rodenrijs, The Netherlands) for stimulating us to work on biological control of poultry red mites. Thea Fiks-Niekerk, Monique Mul and Rick van Emous (Animal Sciences group, Wageningen University, Lelystad, The Netherlands) were instrumental in providing information on pest status of poultry red mites and current practice of poultry management. Sampling of the mite fauna of poultry houses (4 farms in the province of Brabant and Limburg, The Netherlands) was enabled with much help of Nico Harteveld (Koppert BV, Berkel en Rodenrijs, The Netherlands) and Sieds van der Schaaf (Protekta, Gemert, The Netherlands). Ria Wiltenburg and Marnix van de Wetering (CeHaVe, Veghel, The Netherlands) are thanked for their help in realizing the required WOD licence (Wet op de Dierproeven) for tests on blood feeding by mites on young chicken, experiments that were carried out at the research farm 'Laverdonk' (CeHaVe, Heeswijk, The Netherlands). Two M.Sc. students of the University of Groningen, Judith J. Westveer and Linda M. van Zomeren, are thanked for their help with tests on blood feeding by mites on young starlings, experiments that were carried out at Vosbergen estate (Paterswolde, The Netherlands) under WOD licences held by Prof. Jan Komdeur (University of Groningen, Haren, The Netherlands). Tuomi Tuovinen (MTT-Agrifood Research, Jokioinen, Finland) and John Evers (REFONA BV, Westerbork, The Netherlands) are thanked for sharing insights on the use of *Hypoaspis miles* for poultry red mite control. Prof. Urs Wyss (University of Kiel, Germany) is thanked for his skills, patience and help to make video-records of mite behaviour. The work presented in this article was enabled by financial support from the Netherlands Academy of Sciences (KNAW, Amsterdam, The Netherlands), the Dutch Ministry of Agriculture, Nature and Food Quality (LNV, Expertise Center, Bennekom, The Netherlands) and the Product Boards for Livestock, Meat and Eggs (PVE, Zoetermeer, The Netherlands).

Appendix

To test for conspecificity of *Dermanyssus* samples from starling nests and those from poultry farms, three gene regions were sequenced and aligned with homologous sequences from populations of various origins and diverse *Dermanyssus* species: (1) a fragment of 18S–28S rRNA, including internal transcribed spacer 1 (ITS1), 5.8S rRNA and ITS2 (nuclear gene region) (2) 16S rRNA (mitochondrial gene region), and (3) mt-COI (a mitochondrial protein-coding region of cytochrome oxidase subunit I). Populations collected in The Netherlands are labelled by an acronym (IL) referring to the first author of this article, followed by the number of the starling's nest box at Vosbergen estate (Eelde-Paterswolde, The Netherlands). The EMBL Accession numbers are provided below to enable citation of database entries.

Gene region	Population	EMBL accession number
ITS	IL213	FM207490
ITS	IL227	FM207491
16S rRNA	IL213	FM207492
16S rRNA	IL227	FM207494
mt-COI	IL302	FM207495
mt-COI	IL227	FM207496
mt-COI	IL202A	FM207497
mt-COI	IL202C	FM207498
mt-COI	IL213	FM207499

On the basis of mt-COI, Roy et al. (2009) show that four populations (IL302, IL227, IL202, IL213) sampled from starling nests belong to *D. gallinae*. Roy et al. (2008) show that the species delineation apparent from the mt-COI tree is confirmed by other analyses, which include all three gene regions (two mitochondrial and one nuclear). Because the samples from starling populations branch within the most distal *D. gallinae* clade (and neither in a sister clade, nor at the basis of the large *D. gallinae* clade), we infer that they are conspecific. Most likely, all 29 populations from starling-occupied nest boxes at Vosbergen estate in 2007 are conspecific, because Roy et al. (2009) found strong evidence for single or very similar haplotypes of *Dermanyssus* per bird host and geographical location.

Finally, Roy et al. (2009) show that in contrast to others the clade *D. gallinae* groups together all populations represented in poultry farms (layer hens) or breeding facilities for canaries, other Fringillidae and pigeons. Thus, the clade *D. gallinae* harbours synanthropic, as well as bird-associated populations.

References

Abo-Taka SM (1996) Mites inhabiting poultry farms in Egypt. In: Mitchell R, Horn DJ, Needham GR, Welbourn WC (eds) Acarology IX, proceedings, vol 1. Ohio Biological Survey, Columbus, pp 97–99

Axtell RC, Arends JJ (1990) Ecology and management of arthropods pests of poultry. Annu Rev Entomol 35:101–126. doi:10.1146/annurev.en.35.010190.000533

Barker PS (1968) Bionomics of *Androlaelaps casalis* (Berlese) (Acarina: Laelapidae), a predator of mite pests of stored cereals. Can J Zool 46:1099–1102. doi:10.1139/z68-157

Beck W (1999) Farm animals as disease vectors of parasitic epizoonoses and zoophilic dermatophytes and their importance in dermatology. Hautarzt 50:621–628. doi:10.1007/s001050050971

Bloszyk J, Gwiazdowicz DJ, Bajerlein D, Halliday RB (2005) Nests of the white stork *Ciconia ciconia* (L.) as a habitat for mesostigmatic mites (Acari, Mesostigmata). Acta Parasitol 50(2):171–175

Brady J (1970a) The mites of poultry litter: observations on the bionomics of common species, with a species list for England and Wales. J Appl Ecol 7:331–348. doi:10.2307/2401384

Brady J (1970b) Litter mites and their effects on poultry. Worlds Poult Sci J 26:658–668. doi:10.1079/WPS19700022

Brouwer L, Komdeur J (2004) Green nesting material has a function in mate attraction in the European starling. Anim Behav 67:539–548. doi:10.1016/j.anbehav.2003.07.005

Buffoni G, Di Cola G, Baumgärtner J, Maurer V (1995) A mathematical model for trophic interactions in an acarine predator–prey system. J Biol Syst 3:303–312. doi:10.1142/S0218339095000289

Buffoni G, Di Cola G, Baumgärtner J, Maurer V (1997) The local dynamics of acarine predator-prey (*Cheyletus eruditus–Dermanyssus gallinae*) populations: identification of a lumped parameter model. Mitt Schweiz Entomol Ges 70:345–359

Burtt EH Jr, Chow W, Babitt GA (1991) Occurrence and demography of mites of tree swallow, house wren, and eastern bluebird boxes. In: Loye JE, Zuk M (eds) Bird-parasite interactions: ecology, evolution and behaviour. Oxford University Press, Oxford, pp 104–122 (ornithology Series)

Clark L (1991) The nest-protection hypothesis: the adaptive use of plant secondary compounds by European starlings. In: Loye JE, Zuk M (eds) Bird–parasite interactions: ecology, evolution and behaviour. Oxford University Press, Oxford, pp 205–221 (ornithology Series)

El-Kammah KM, Oyoun LMI (2007) Taxonomic records of mite species associated with animal and poultry farms in the Nile Delta, valley and N. Sinai Egypt. In: Morales-Malacara JB, Behan-Pelletier V, Ueckermann E, Perez-T, Estrada-Venegas EG, Badii M (eds) Acarology XI, proceedings of the international congress, pp 223–232

Emous RA, Fiks-Van Niekerk TGCM, Mul MF (2005) Enquete vogelmijten op leghennenbedrijven: € 11. 000.000 schade voor de sector. Pluimveehouderij 35(22 januari):8–9

Evans GO (1957) An introduction to the British Mesostigmata (Acarina) with keys to families and genera. Zool J Linn Soc 43:203–259. doi:10.1111/j.1096-3642.1957.tb01552.x

Faraji F, Bakker FM (2008) A modified method for clearing, staining and mounting plant-inhabiting mites. Eur J Entomol 105:793–795

Fenda P, Lengyel J (2007) Rostoce (Acarina: Mesostigmata) v hniezdach orliaka morskeho (*Haliaeetus albicilla*) na Slovensku. Entomofauna carpathica 19:48–50

Gupta SK, Paul K (1989) Nest-associated acarines of birds in India. In: Channabasavanna GP, Viraktamath CA (eds) Progress in acarology, vol 2. EJ Brill, Leiden, The Netherlands, pp 315–321

Gwiazdowicz DJ (2003) Mites (Acari: Mesostigmata) occurring in the nests of birds of prey (Falconiformes) and owls (Strigiformes). Acarina 11:235–239

Gwiazdowicz DJ, Mizera T, Skorupski M (1999) Mites from greater spotted eagle nests. J Raptor Res 33:257–260

Gwinner H, Berger S (2005) European starlings: nestling condition, parasites and green nest material during the breeding season. J Ornithol 146:365–371. doi:10.1007/s10336-005-0012-x

Hughes AM (1976) The Mites of stored food and houses. Second Edition, Ministry of Agriculture, Fisheries and Food, Technical Bulletin 9:1–400

Kristofík J, Masan P, Sustek Z (1996) Ectoparasites of bee-eater (*Merops apiaster*) and arthropods in its nests. Biologia 51:557–570

Kristofík J, Masan P, Sustek Z (2001) Mites (Acari), beetles (Coleoptera) and fleas (Siphonaptera) in the nests of great reed warbler (*Acrocephalus arundinaceus*) and reed warbler (*A. scirpaceus*). Biologia 56:525–536

Kristofík J, Masan P, Sustek Z, Kloubec B (2003) Arthropods (Pseudoscorpionida, Acari, Coleoptera, Siphonaptera) in nests of the Tengmalm's owl, *Aegolius funereus*. Biologia 58:231–240

Lesna I, Sabelis MW, Bolland HR, Conijn CGM (1995) Candidate natural enemies for control of *Rhizoglyphus robini* Claparède (Acari: Astigmata) in lily bulbs: exploration in the field and preselection in the laboratory. Exp Appl Acarol 19:655–669. doi:10.1007/BF00145254

Li G, Meng Y (1992) Investigation in sucking mice free blood habit of *Hypoaspis lubrica*. Chin J Vector Biol Control 3:9–12 (in Chinese)

Lundqvist L (1995) Ectoparasitic mites in the nest of starlings (*Sturnus vulgaris*); do they play a role in the sexual selection of the birds? In: Kropczynska D, Boczek J, Tomczyk A (eds) The Acari: physiological and ecological aspects of Acari-host relationships. Oficyna Dabor, Warszawa, Poland, pp 545–548

Maurer V, Baumgärtner J (1992) Temperature influence on life table statistics of the chicken mite *Dermanyssus gallinae* (Acari: Dermanyssidae). Exp Appl Acarol 15:27–40. doi:10.1007/BF01193965

Maurer V, Baumgärtner J (1994) A population model for *Dermanyssus gallinae* (Acari, Dermanyssidae). Exp Appl Acarol 18:409–422. doi:10.1007/BF00051523

Maurer V, Hertzberg H (2001) Ökologische legehennenhaltung. Was tun gegen die kleinen Vampire? DGS-Magazin. Woche 40:49–52

Maurer V, Baumgärtner J, Bieri M, Fölsch DW (1993) The occurrence of the chicken mite *Dermanyssus gallinae* (Acari: Dermanyssidae) in Swiss poultry houses. Mitt Schweiz Entomol Ges 66:87–97

Mazgajski TD (2007) Effect of old nest material in nest boxes on ectoparasite abundance and reproductive output in the European starling *Sturnus vulgaris* (L.). Pol J Ecol 55:377–385

McKinley DJ (1963) The morphology and biology of *Haemolaelaps casalis* Berlese (Acarina: Mesostigmata). Ann Mag Nat Hist 13(6):65–76

Men YT (1959) Concerning the feeding of the mite *Haemolaelaps casalis* (Gamasoidea, Parasitiformes). Medskaya Parazitol 28:603–609 (in Russian)

Nordenfors H, Chirico J (2001) Evaluation of a Sampling Trap for *Dermanyssus gallinae* (Acari: Dermanyssidae). J Econ Entomol 94:1617–1621

Pacejka AJ, Thompson CF (1996) Does removal of old nests from nest boxes by researchers affect mite populations in subsequent nests of house wrens? J Field Ornithol 67:558–564

Pacejka AJ, Santana E, Harper RJ, Thompson CF (1996) House wrens *Troglodytes aedon* and nest-dwelling ectoparasites: mite population growth and feeding patterns. J Avian Biol 27:273–278. doi:10.2307/3677258

Pacejka AJ, Gratton CM, Thompson CF (1998) Do potentially virulent mites affect house wren (*Troglodytes aedon*) reproductive success? Ecology 79:1797–1806

Philips JR (2000) Review and checklist of the parasitic mites (Acarina) of the Falconiformes and Strigiformes. J Raptor Res 34:210–231

Philips JR, Dindal DL (1979) The acarine community of nests of birds of prey. In: Rodriguez JG (ed) Recent advances of acarology, vol 1. Academic Press, New York, pp 559–562

Philips JR, Poole A, Holt D (1989) Nest mites of ospreys and short-eared owls in Massachusetts salt marshes, USA. In: Channabasavanna GP, Viraktamath CA (eds) Progress in acarology, vol 2. EJ Brill, Leiden, The Netherlands, pp 305–307

Proctor H, Owens I (2000) Mites and birds: diversity, parasitism and coevolution. Trends Ecol Evol 15:358–364. doi:10.1016/S0169-5347(00)01924-8

Pung OJ, Carlile LD, Whitlock J, Vives SP, Durden LA, Spadgenske E (2000) Survey and host fitness effects of red-cockaded woodpecker blood parasites and nest cavity arthropods. J Parasitol 86:506–510

Putatunda BN, Gupta SK, Singh J (1989) Acarine associates of birds in West Bengal, India. In: Channabasavanna GP, Viraktamath CA (eds) Progress in acarology, vol 2. EJ Brill, Leiden, The Netherlands, pp 309–313

Quintero MT, Acevedo AC (1984) Studies on deep litter mites in poultry farms in Mexico. In: Griffiths DA, Bowman CE (eds) Acarology VI, vol 1. Ellis Horwood Ltd, Chichester, pp 629–634

Rabash J, Browne W, Healy M, Cameron B, Charlton C (2008) "MLwiN v2.10 beta 5" Centre of Multilevel Modelling. University of Bristol, UK

Radovsky FJ (1985) Evolution of mammalian mesostigmate mites. In: Kim KC (ed) Coevolution of parasitic arthropods and mammals. Wiley, New York, pp 441–504

Radovsky FJ (1994) The evolution of parasitism and the distribution of some dermanyssoid mites (Mesostigmata) on vertebrate hosts. In: Houck MA (ed) Mites: ecological and evolutionary analysis of life history patterns. Chapman and Hall, New York, pp 186–217

Rosen S, Yeruham I, Braverman Y (2002) Dermatitis in humans associated with the mites *Pyemotes tritici, Dermanyssus gallinae, Ornithonyssus bacoti* and *Androlaelaps casalis* in Israel. Med Vet Entomol 16:442–444. doi:10.1046/j.1365-2915.2002.00386.x

Roy L, Chauve CM (2007) Historical review of the genus *Dermanyssus* Dugès (Acari: Mesostigmata: Dermanyssidae). Parasite 14:87–100

Roy L, Dowling APG, Chauve CM, Buronfosse T (2008) Delimiting species boundaries within *Dermanyssus* Dugès, 1834 (Acari: Mesostigmata) using a total evidence approach. Mol Phylogenet Evol (in press). doi:10.1016/j.ympev.2008.11.012

Roy L, Dowling APG, Chauve CM, Lesna I, Sabelis MW, Buronfosse T (2009) Molecular phylogenetic assessment of host range in *Dermanyssus* species. Exp Appl Acarol. doi:10.1007/s10493-008-9231-1

Svana M, Fenda P, Orszaghova Z (2006) Rostoce (Acarina: Mesostigmata) v hniezdach vtakov jz Slovenska. Folia faunistica Slovaka 11:39–42

Tenquist JD, Charleston WAG (2001) A revision of the annotated checklist of ectoparasites of terrestrial mammals in New Zealand. J R Soc N Z 31:481–542

Tuovinen T (2008) Predatory arthropods as biocontrol agents of *Dermanyssus gallinae*? In: BSP Spring, trypanosomiasis/leishmaniasis & malaria meetings. Newcastle, UK, March 30–April 2. Abstract booklet: 170

Valiente Moro C, Chauve C, Zenner L (2005) Vectorial role of some dermanyssoid mites (Acari, Mesostigmata, Dermanyssoidea). Parasite 12:99–109

Valiente Moro C, Chauve C, Zenner L (2007) Experimental infection of *Salmonella Enteritidis* by the poultry red mite, *Dermanyssus gallinae*. Vet Parasitol 146:329–336. doi:10.1016/j.vetpar.2007.02.024

Veiga JP, Polo V, Vinuela J (2006) Nest green plants as a male status signal and courtship display in the spotless starling. Ethology 112:196–204. doi:10.1111/j.1439-0310.2006.01148.x

Xing-Yuan M, Xian-Guo G, Wen-Ge D, Ai-Qin N, Ti-Jun Q, Dian W (2007) Ectoparasites of Chevrier 's field mouse, *Apodemus chevrieri*, in a focus of plague in Southwest China. Med Vet Entomol 21:297–300

The testing of antibodies raised against poultry red mite antigens in an in vitro feeding assay; preliminary screen for vaccine candidates

Harry W. Wright · Kathryn Bartley · Alasdair J. Nisbet · Regina M. McDevitt · Nickolas H. C. Sparks · Sarah Brocklehurst · John F. Huntley

Originally published in the journal Experimental and Applied Acarology, Volume 48, Nos 1–2, 81–91.
DOI: 10.1007/s10493-009-9243-5

Abstract *Dermanyssus gallinae* (De Geer), the poultry red mite, is a blood-feeding ectoparasite that infests many bird species. We have used an in vitro feeding assay to allow the identification of protective *D. gallinae* antigens that may have potential as vaccine candidates. Homogenised mites were extracted sequentially with PBS, Tween 20, Triton X100 and urea giving four protein fractions. Five experimental groups of Lohmann Brown hens were used to generate antibodies; four groups were injected with one of each of the protein fractions in QuilA adjuvant and a control group was injected with adjuvant only. Booster injections were administered 2 and 4 weeks after initial immunisation. Eggs were collected throughout the experiment and soluble IgY antibodies were extracted from a pool of egg yolks collected at week six post-injection. Western blots, performed using post vaccination antibodies from test and control groups, revealed a strong antibody response against a range of injected proteins. Fresh chicken blood, supplemented with antibodies raised against these protein fractions, was fed to mites in an in vitro feeding assay in order to determine whether the antibodies had an anti-mite effect. Although there was variability in the numbers of feeding mites, it was found that the strongest anti-mite effect was seen with the PBS protein fraction, which had a cumulative average mortality of 34.8% 14 days after feeding compared with 27.3% for the control group ($P = 0.043$).

H. W. Wright (✉) · K. Bartley · A. J. Nisbet · J. F. Huntley
Parasitology Division, Moredun Research Institute, Midlothian EH26 0PZ, Scotland, UK
e-mail: Harry.Wright@moredun.ac.uk

R. M. McDevitt
Public Health Department, NHS Ayrshire & Arran, Afton House, Dalmellington Road, Ayr KA6 6AB, UK

N. H. C. Sparks
Avian Science Research Centre, Animal Health Group, SAC Ayr, Auchincruive KA6 5HW, Scotland, UK

S. Brocklehurst
Biomathematics & Statistics Scotland, James Clerk Maxwell Building, King's Buildings, Mayfield Road, Edinburgh EH9 3JZ, Scotland, UK

Keywords *Dermanyssus gallinae* · Poultry red mite · Vaccine · Protective antigens

Introduction

Dermanyssus gallinae is a haematophagous parasite that infests many bird species. From an economic point of view it is one of the most important ectoparasites affecting egg-laying chickens in many countries (Chauve 1998). A recent survey carried out in Northern England showed that 87.5% of farms responding to a survey were infested with *D. gallinae* (Guy et al. 2004). In Sweden, 67% of backyard flocks were found to be infested with *D. gallinae* (Hoglund et al. 1995). *D. gallinae* rest during the day in any dark place around the framework of machinery and nest boxes of the poultry house. The mites infest birds and feed for approximately an hour and a half during the night before retreating to these dark hiding places to digest their blood meal after which they can moult, reproduce and lay eggs. Mites can survive for up to 8 months under natural conditions without a blood meal (Kirkwood 1963). The ability of *D. gallinae* to survive for such long periods without a blood meal makes the de-infestation of barns and cages a problem that needs to be addressed before a new flock is introduced, especially in a commercial environment where rapid turnover is a priority. A fallow period of at least 4 weeks between housing separate flocks has been suggested to allow the mite population to decrease (Beugnet et al. 1997), but this is not viewed by the egg producing industry as commercially viable.

Conventional mite control methods involve spraying of cages and housing with chemical acaricides. However, this is not always effective because mites in cracks and crevices escape acaricide treatment and resistance to pyrethroids has been reported in a number of European countries (Beugnet et al. 1997; Nordenfors et al. 2001).

Therefore a number of alternative mite control methods are being explored and vaccination is a potential mite control strategy. During a natural infestation by parasites the host is continually being exposed to antigens (known as "exposed" antigens) from the parasite (Willadsen and Kemp 1988). In the case of *D. gallinae*, however, the natural immune response to exposed antigens appears ineffective in controlling mite populations in poultry houses. A previous study has shown that there was no correlation between mite populations and IgY antibody titres in egg yolk (Arkle et al. 2006) and Western blots of *D. gallinae* soluble protein fraction probed with IgY extracted from egg yolks collected from a commercial farm infested with mites did not show substantial detection of any antigens (Bartley et al. 2008).

The alternative to using "exposed" antigens as vaccine candidates is to exploit "concealed" antigens. These are internal proteins of the parasite which are not normally exposed to the host or it's immune system. The commercialised tropical cattle tick vaccine TickGARD is based on a concealed gut membrane-associated glycoprotein (Bm86) and generates a protective antibody response against the tick *Rhipicephalus (Boophilus) microplus*. Ingested antibodies bind to Bm86 in the tick gut and cause damage leading to a high tick mortality as well as a reduction in the egg laying ability and the survival of the eggs laid by ticks feeding on vaccinated cattle (Rand et al. 1989; Willadsen et al. 1989).

The purpose of the work described here was to test the efficacy of mite proteins in eliciting a protective antibody response by testing antibodies raised against a number of mite antigens using an in vitro feeding assay. The proteins of interest would probably be concealed due to the lack of evidence to suggest a natural immunity developed against exposed antigen, as previously mentioned. Feeding assays for *D. gallinae* have been developed previously (Kirkwood 1971; McDevitt et al. 2006), and a modification of the McDevitt et al. techniques was employed in the current study. This has enabled the preliminary testing of antibodies raised against *D. gallinae* proteins to determine potential vaccine candidates, before embarking on the more technically demanding and expensive process of in vivo challenge studies. While the latter will be required to validate any potential vaccine, the current in vitro studies will allow the rapid selection of effective vaccine candidates, and enable studies on their mode(s) of action.

Materials and methods

Dermanyssus gallinae used in the experiments

Mites were collected from a commercial egg production unit and stored in a 75 cm^2 culture flask (Corning), with a vented cap. These mites were conditioned before use in the in vitro feeding assay by initial incubation at room temperature ($\sim$21 ± 2°C) for 7 days, followed by storage at 4 ± 1°C for 21 days (McDevitt et al. 2006).

Within 4 h of collection, mites, which were a mix of life stages and the majority (>60%) of the mites were engorged, were taken from the culture flask and snap frozen in liquid nitrogen and stored at −80°C before extraction and fractionation of proteins.

Fractionation of *Dermanyssus gallinae* proteins

Mite proteins were fractionated by homogenising 1 g of mites and extracting sequentially with: (1) phosphate buffered saline (PBS, soluble proteins), (2) PBS containing 0.1% v/v Tween 20 (membrane associated proteins), (3) PBS containing 2% v/v Triton X100 (integral membrane proteins) and (4) 10 M urea containing 2 M dithiothreitol [(DTT), insoluble proteins]. Frozen mites were added to 10 ml of chilled PBS and homogenised using a mechanical homogeniser (Ultra-Turrax T25, IKA-Labortechnik) for 30 s, on the highest power setting, rested on ice for 1 min then homogenised for a further 30 s. The homogenised mites were then centrifuged at 25,000*g* for 20 min at 4°C. The supernatant (PBS extract) was retained on ice and the pellet was re-suspended in chilled PBS containing 0.1% v/v Tween 20 and transferred to a Dounce homogeniser, homogenised then centrifuged as previously described. This process was repeated using a further aliquot of PBS containing 0.1% v/v Tween 20. The supernatant was retained on ice and the pellet was resuspended in PBS containing 2% v/v Triton X100 before centrifugation at 100,000*g* for 45 min at 4°C. The supernatant was retained and the pellet re-suspended in 10 M urea and 2 M DTT and incubated at 37°C overnight. The tubes were then centrifuged at 100,000*g* for 20 min at room temperature. All supernatants were stored at −20°C.

Immunisation of hens with the four protein fractions

An immunisation study was carried out to raise antibodies against the four protein fractions. Forty Lohmann Brown hens were sourced at 16 weeks of age and housed in a single room in ten floor pens in groups of four under standard environmental and feeding conditions for 4 weeks prior to the study commencing. The hens matured and began egg production at 19–20 weeks of age. At 20 weeks of age, the hens were divided into 10 groups of four on the basis of body weight, ensuring that the mean weights for each group were within a 100 g tolerance and placed in floor pens under standard environmental and feeding conditions. Two pens of hens were randomly assigned to each treatment. The hens from each treatment group were immunised with one of the four mite extracts or adjuvant control. Each dose comprised 75 μg of mite protein (determined using the Pierce BCA protein assay for soluble protein, membrane associated protein and integral membrane protein fractions; the 2-D Quant Kit (Amersham Biosciences) was used to determine the concentration of the insoluble protein fraction), 200 μg of QuilA adjuvant, and PBS in a final volume of 300 μl. Control immunisations consisted of 200 μg of QuilA adjuvant in PBS. The injections were administered into alternate breast muscles of the hen 1 week after they had been assigned to groups. Following the primary injection, two booster injections were administered at 2 week intervals.

The experiment was carried out in accordance with the Animals (Scientific Procedures) Act 1986. The experiment was ratified by the ethics committee at the Scottish Agricultural College in Auchincruive.

PAGE gel and Western blots

Thirteen microlitres of each protein fraction (3.5 μg) were separated by electrophoresis on a 4–12% NuPAGE gel (Invitrogen) under reducing conditions using MES buffer, following the manufacturer's instructions. To monitor the hens' antibody response to the mite extracts, Western blots were prepared using the Invitrogen iBlot system to transfer the proteins onto nitrocellulose membranes following the manufacturer's instructions. The primary antibodies used in the Western blots were extracted from eggs collected 6 weeks after initial injection. Briefly, one gram of egg yolk, taken from a pooled sample, was homogenised in 10 ml of PBS, incubated at 37°C for 5 min and vortexed for 5 min. The egg yolk debris was pelleted by centrifugation for 5 min at 3,000*g* and the supernatant was stored at −20°C before use. The supernatant containing IgY antibody was used, without further dilution, as the primary antibody. The nitrocellulose blot was cut into strips and the strips were incubated in primary antibody overnight at 4°C on an orbital rocker and then washed in wash buffer (PBS containing 0.5 M NaCl, 0.5% Tween 80) changing every 20 min for 3 h. The secondary antibody was rabbit anti-chicken IgY horse radish peroxidase conjugate (Sigma), diluted 1:30000 in wash buffer before use. The blot was incubated in secondary antibody at room temperature for 3 h on an orbital rocker before being washed every 20 min with wash buffer for 3 h. The blots were developed using 3,3′-diaminobenzidine (DAB, Sigma).

Antibody extraction from egg yolk for in vitro assay

The eggs used were collected 6 weeks after initial injection and the yolks were pooled and stored at −20°C. Egg yolk (0.5 g) was added to PBS (750 μl) and incubated at 37°C for 10 min. This was then mixed thoroughly for 5 min before centrifugation at 30,000*g*. The supernatant, containing the antibodies, was decanted and stored at −20°C until required.

In vitro feeding assay

This method was based on a protocol described previously (McDevitt et al. 2006) and modified as follows: The bases of 5 ml cryovial tubes (Nalgene) were cut off and smoothed down. Plastic pastettes, cut at the half way point of the bulb and ca. 2 cm up the stalk, were inverted and used as a blood reservoir and to hold the chick skin in place (Fig. 1). Thoroughly washed skins from 1-day-old chicks, that had been plucked and harvested previously and stored in ethanol (100%) at 4°C for 2 weeks, were cut into 2 cm^2 pieces and placed over the end of the tube with the external surface of the skin facing the mites. The skin was held in place by the blood reservoir. The filter paper, cut into strips 40 mm long and 5 mm wide, was used to gather mites and placed in the prepared chamber. The screw cap of the chamber was then attached.

Extracted antibody (50 μl) was added to 950 μl of fresh heparinised chicken blood (36 USP-units heparin sodium salt/ml blood) and mixed by inversion. The antibody titres were at approximately physiological conditions. Approximately 250 μl of blood was then added to the blood reservoir and the chambers were incubated overnight at 40°C in darkness at a relative humidity of 75% for between 18 and 24 h. Eight feeding chambers were assembled per treatment group, where each group represented each mite protein extract or adjuvant control. The experiment was repeated 3 times. If the mites had fed within the past 24 h, i.e. had signs of fresh blood in their gut, they were placed into individual wells of a 96 well plate (Greiner bio-one). The mortality of the mites was monitored daily over a 3 week period. Differences in mite mortality over the whole experiment between the four mite

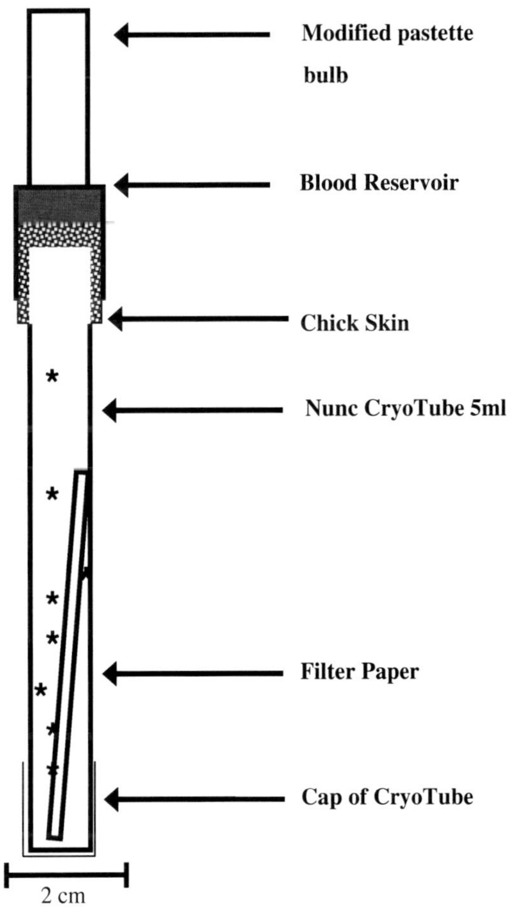

Fig. 1 In vitro feeding device for *Dermanyssus gallinae*, modified from McDevitt et al. (2006)

protein fractions and the control were statistically analysed using a proportional hazards model (Collet 1994), adjusting for the inclusion of data from the three experiments. Differences in the mortality data for the first 2 days of the three experiments between the four mite protein fraction treatment groups and the control group were analysed using a general linear mixed model with binomial error and logit link function, fitting treatment group as a fixed effect and experiment as a random effect. Statistical analyses were carried out using Genstat (Genstat, 10th edition, Lawes Agricultural Trust, VSN International Ltd, Oxford, UK).

Results

Protein fractionation

Each sequential extraction technique yielded a number of proteins with a broad range of molecular masses. The protein profiles of these fractions from the mites are shown in Fig. 2 and were clearly different, showing that the different treatment groups were injected with different proteins.

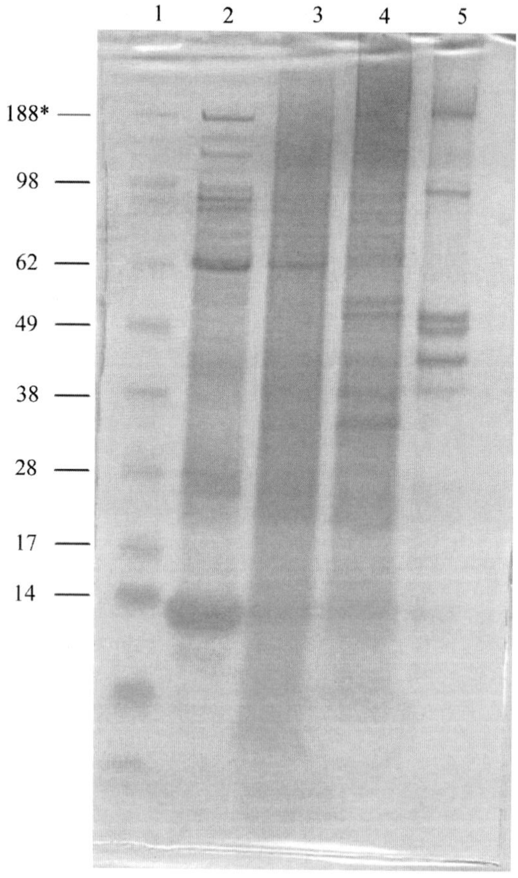

Fig. 2 A polyacrylamide gel showing the protein profiles of the four fractions injected into hens in the vaccine trial, 3.5 μg of each protein was loaded. *Lane 1* SeeBlue plus 2 protein markers (Invitrogen). *Lane 2* PBS soluble proteins. *Lane 3* membrane associated proteins. *Lane 4* integral membrane proteins. *Lane 5* insoluble proteins. The gel was stained with SimplyBlue (Invitrogen) stain. *Molecular mass (kDa)

Immunoreactivity of chickens to protein fractions

A Western blot was performed using soluble IgY extracts of egg yolks from hens vaccinated with the four mite protein fractions as the primary antibody source and 3.5 μg of mite protein fraction immobilised onto nitrocellulose membranes. A number of proteins with a broad range of molecular masses were recognised by IgY from vaccinated hens and the profiles of proteins recognised differed between the mite extracts used for vaccination (Fig. 3). When comparing the reactivity of the vaccinated groups to the reactivity of the adjuvant-only groups it can be seen that there is a marked difference in the number and intensity of the bands. Some of the bands detected in both the vaccinated and control groups represented the two IgY subunits. Bands of ca. 62 and 25 kDa in lanes 1 and 6 and to a lesser extent 2 and 7 most likely represent the IgY subunits present in the mites' blood meal that were detected with the anti-IgY secondary antibody.

The differences between the profile of the SDS–PAGE gel and the banding patterns of the Western blot could be explained by the co purification of chicken proteins present in the mites gut that were injected into the chicken. Because of MHC restriction the chicken would not mount an antibody against "self" proteins. The immunogenic proteins that create bands on the Western blot might not be very abundant.

In vitro feeding assay

The in vitro feeding assay was used to feed mites fresh pooled heparinised blood from Lohmann Brown hens mixed with antibodies extracted from egg yolks of hens vaccinated with the 4 mite protein fractions and the adjuvant-only control vaccination. Because of uncontrollable variability in the numbers of mites feeding in each chamber, various numbers of mites progressed into the mortality monitoring phase of the experiments

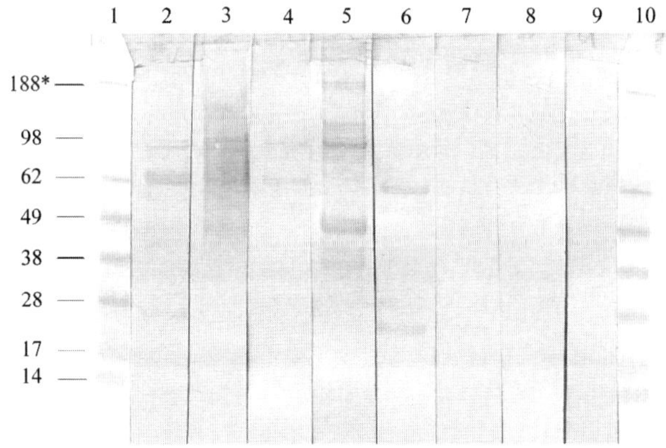

Fig. 3 Western blot of mite (*Dermanyssus gallinae*) proteins (3.5 μg/lane) probed with antibodies extracted from egg yolk derived from immunised chickens. *Lane 2* PBS-soluble proteins, probed with antibodies generated against water soluble proteins. *Lane 3* membrane associated proteins, probed with antibodies generated against membrane associated proteins. *Lane 4* integral membrane proteins, probed with antibodies generated against integral membrane proteins. *Lane 5* insoluble proteins, probed with antibodies generated against insoluble proteins. *Lane 6* PBS-soluble proteins, probed with antibodies generated in the control group. *Lane 7* membrane associated proteins, probed with antibodies generated in the control group. *Lane 8* integral membrane proteins, probed with antibodies generated in the control group. *Lane 8* insoluble proteins, probed with antibodies generated in the control group. *Lanes 1 and 10* SeeBlue plus II protein markers (Invitrogen). *Molecular mass (kDa)

(Table 1). Because the number of mites feeding was variable for different groups and experiments, mortality was expressed as a percentage to allow unbiased summary comparison between treatment groups and each experiment. However, while estimates will be unbiased, the power to test the effect of the four mite protein fractions will vary marginally with the total number of feeding mites per fraction over the three experiments. The percentage mite mortality differed between the four groups injected with mite protein (Fig. 4). The percentage mortality was highest when the mites were fed antibodies raised against the PBS soluble protein fraction, with cumulative percentage mortality (across three experimental repeats) of 16.4% after 24 h, compared to 10.9% for the controls. The cumulative percentage mortality in the PBS soluble treatment rose from 16.4 to 34.8% after 14 days, whereas the cumulative percentage mortality in the control mites rose from 10.9 to 27.3% after 14 days. The proportional hazards model showed that there was a statistically significant difference in mortality rates between the five treatments ($P = 0.043$). Parameter estimates showed that the anti-PBS protein fraction antibodies caused the highest mean mortality compared to the control group but that this difference was not statistically significant.

Focusing on the data from the first 2 days of the experiment (Fig. 4) a difference was observed between the gradient, or mortality rates of the PBS group and control group, with the rate for the PBS group consistently greater than that for the control group. This, however, was not the case from day three onwards. The generalised linear mixed model showed that there was a significant difference in the mean mortality in the first 2 days between the five groups ($P = 0.016$). Parameter estimates showed that the anti-PBS antibodies caused the highest mean mortality compared to the control and that this effect was statistically significant ($P = 0.013$).

Discussion

Here we have demonstrated that when mite-specific antibodies were raised against *D. gallinae* extracts and tested in an in vitro feeding assay, the antibodies generated against the PBS-soluble mite proteins resulted in higher mite mortality than in control groups. The protective, anti-mite, antibodies were raised against proteins from this fraction that have not yet been identified. During the first 2 days after feeding the mean mortality of the mites from the PBS group was significantly higher than that of the control group. This mortality may be attributable to antibodies directed towards a single antigen, or to multiple antigens, important to the normal functioning and viability of the mite. These antigens appear to be present predominantly in the PBS soluble fraction, and these soluble proteins therefore represent a likely source of vaccine candidate(s) which require identification and characterisation. The increased mortality of fed mites was modest but it should be noted that, during the development of TickGARD, which is an effective vaccine against ticks that has

Table 1 The number of fed mites monitored in each group for the three experiments

Group	Experiment 1	Experiment 2	Experiment 3
Control	98	24	43
PBS soluble	58	100	92
Membrane associated	57	72	61
Integral membrane	104	24	28
Insoluble	78	189	51

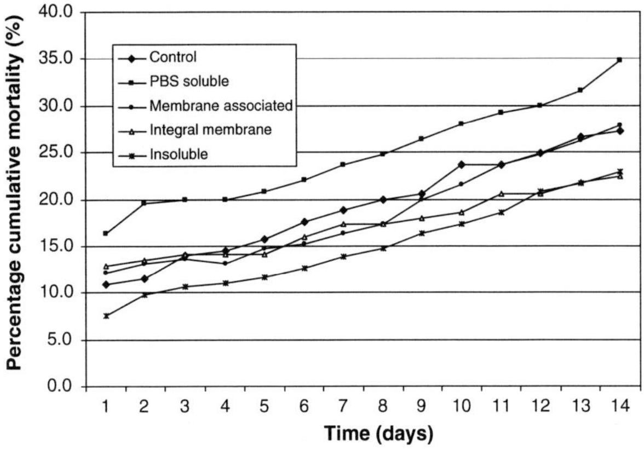

Fig. 4 Cumulative percentage mortality of *Dermanyssus gallinae* fed heparinised chicken blood enriched with antibodies raised against four mite protein fractions and adjuvant-only. Mortality was monitored daily over a 14 day period following feeding. Each point represents cumulative mean percentage mortality and is derived from data recorded in 3 independent experiments. The range of the Standard Error of the mean were Control 0.55–6.18; PBS-soluble group 3.36–8.04; Membrane associated group 1.80–6.22; Integral membrane group 0.93–5.00; Insoluble group 2.54–3.96

been commercialised, the initial effects of vaccination with crude tick extracts were also comparatively modest. As the protein component of the vaccine was reduced in complexity, the efficacy increased up to 40% (Johnston et al. 1986), and to 90% by use of either pure native or recombinant Bm86 (Willadsen et al. 1995). It is also notable that the increased mortality seen in our experiments resulted after a single blood meal, whereas in nature the mites would feed repeatedly, potentially generating an enhanced cumulative effect of the PBS-soluble fraction on mortality.

The numbers of mites feeding was variable. In some of the chambers up to 50% of the mites fed whereas in others none were observed to feed. This may have depended upon the part of the body from which the skin was derived, and factors such as skin thickness or attached mesoderm may have influenced the mite's ability to penetrate and feed. In *D. gallinae* a kairomone has been identified from feeding mites that attracts more mites to the site of feeding (Zeman 1988). This kairomone may have differed in concentration in groups of mites feeding in separate chambers, affecting the overall feeding of mites in each replicate.

Future work to identify vaccine candidates will concentrate on antibodies raised against the PBS soluble proteins of poultry red mites as this appears to be the most promising extract. The efficacy of PBS-soluble proteins as vaccine candidates against mites differs somewhat from results for the tick *R. microplus* where membrane-associated proteins gave the most promising results in reducing populations as discussed above (Willadsen et al. 1995). The data presented here are, however, in agreement with work performed previously, using a similar approach to identify protective antigens from *P. ovis*, where PBS-soluble components were more protective than any other extracts, eliciting a 15-fold reduction in mite burden and four-fold reduction in scab lesion when sub-fractionated PBS extracts were used in in vivo vaccine trials (Smith et al. 2002).

Currently, the most successful control of *D. gallinae* is based on cage cleaning and acaricide spraying of empty cages before restocking (Chauve 1998). Only one acaricide spray ByeMite®) is licensed for use when cages are stocked. One of the main problems with the use of chemical control is the emergence of acaricide resistant mites. Proteins which would be used in vaccine preparations have multiple epitopes where the antibodies can bind, so there would have to be multiple mutations in the DNA encoding these epitopes, changing the amino acid sequence and configuration for any parasite to become resistant to a protein vaccine. Such an event would more than likely change the function of the protein and take a long period of time to occur. Hence the development of resistance to a vaccine based on targeting "hidden" or "exposed" antigens is highly unlikely.

Studies are being planned to sub-fractionate the PBS fraction to generate a more focussed antibody response in the host and thus challenge the mites with antibodies targeting a less complex mix of protective proteins. Once identified, recombinant versions of these proteins will be tested in an in vivo trial.

In summary we have presented preliminary data indicating the protective capability of soluble mite proteins and the potential of vaccination as a method of controlling *D. gallinae* populations.

Acknowledgments We gratefully acknowledge funding by the British Egg Marketing Board Educational Trust and DEFRA. We would also like to thank Dr. Stuart Smith (Moredun) for his help with protein purification and local poultry farmers for their assistance.

References

Arkle S, Guy JH, Sparagano O (2006) Immunological effects and productivity variation of red mite (*Dermanyssus gallinae*) on laying hens- implications for egg production and quality. Worlds Poult Sci J 62:249–257. doi:10.1079/WPS200594

Bartley K, Nisbet AJ, Offer JE, Sparks NH, Wright HW, Huntley JF (2008) Histamine release factor from *Dermanyssus gallinae* (De Geer): characterization and in vitro assessment as a protective antigen. Int J Parasitol. doi:10.1016/j.ijpara.2008.09.006

Beugnet F, Chauve C, Gauthey M, Beert L (1997) Resistance of the red poultry mite to pyrethroids in France. Vet Rec 140:577–579

Chauve C (1998) The poultry red mite *Dermanyssus gallinae* (De Geer, 1778): current situation and future prospects for control. Vet Parasitol 79:239–245. doi:10.1016/S0304-4017(98)00167-8

Guy JH, Khajavi M, Hlalel MM, Sparagano O (2004) Red mite (*Dermanyssus gallinae*) prevalence in laying units in Northern England. Br Poult Sci 45(suppl 1):S15–S16. doi:10.1080/00071660410001698001

Hoglund J, Nordenfors H, Uggla A (1995) Prevalence of the poultry red mite, *Dermanyssus gallinae*, in different types of production systems for egg layers in Sweden. Poult Sci 74:1793–1798

Johnston LA, Kemp DH, Pearson RD (1986) Immunization of cattle against *Boophilus microplus* using extracts derived from adult female ticks: effects of induced immunity on tick populations. Int J Parasitol 16:27–34. doi:10.1016/0020-7519(86)90061-5

Kirkwood AC (1963) Longevity of the mites *Dermanyssus gallinae* and *Liponyssus sylviarum*. Exp Parasitol 14:358–366. doi:10.1016/0014-4894(63)90043-2

Kirkwood AC (1971) In vitro feeding of *Dermanyssus gallinae*. Exp Parasitol 29:1–6. doi:10.1016/0014-4894(71)90002-6

McDevitt R, Nisbet AJ, Huntley JF (2006) Ability of a proteinase inhibitor mixture to kill poultry red mite, *Dermanyssus gallinae* in an in vitro feeding system. Vet Parasitol 141:380–385. doi:10.1016/j.vetpar.2006.05.013

Nordenfors H, Hoglund J, Tauson R, Chirico J (2001) Effect of permethrin impregnated plastic strips on *Dermanyssus gallinae* in loose-housing systems for laying hens. Vet Parasitol 102:121–131. doi:10.1016/S0304-4017(01)00528-3

Rand KN, Moore T, Sriskantha A, Spring K, Tellam R, Willadsen P, Cobon GS (1989) Cloning and expression of a protective antigen from the cattle tick *Boophilus microplus*. Proc Natl Acad Sci USA 86:9657–9661. doi:10.1073/pnas.86.24.9657

Smith WD, Bates P, Pettit DM, Van den Broek A, Taylor MA (2002) Attempts to immunize sheep against the scab mite, *Psoroptes ovis*. Parasite Immunol 24:303–310. doi:10.1046/j.1365-3024.2002.00469.x

Willadsen P, Kemp DH (1988) Vaccination with 'concealed' antigens for tick control. Parasitol Today 4:196–198. doi:10.1016/0169-4758(88)90084-1

Willadsen P, Riding GA, McKenna RV, Kemp DH, Tellam RL, Nielsen JN, Lahnstein J, Cobon GS, Gough JM (1989) Immunologic control of a parasitic arthropod. Identification of a protective antigen from Boophilus microplus. J Immunol 143:1346–1351

Willadsen P, Bird P, Cobon GS, Hungerford J (1995) Commercialisation of a recombinant vaccine against *Boophilus microplus*. Parasitology 110:S43–S50

Zeman P (1988) Surface skin lipids of birds—a proper host kairomone and feeding inducer in the poultry red mite, *Dermanyssus gallinae*. Exp Appl Acarol 5:163–173. doi:10.1007/BF02053825

The poultry red mite (*Dermanyssus gallinae*): a potential vector of pathogenic agents

Claire Valiente Moro · Carlos J. De Luna · Alexander Tod · Jonathan H. Guy · Olivier A. E. Sparagano · Lionel Zenner

Originally published in the journal Experimental and Applied Acarology, Volume 48, Nos 1–2, 93–104.
DOI: 10.1007/s10493-009-9248-0

Abstract The poultry red mite, *D. gallinae* has been involved in the transmission of many pathogenic agents, responsible for serious diseases both in animals and humans. Nowadays, few effective methods are available to control the ectoparasite in poultry farms. Consequently, this is an emerging problem which must be taken into account to maintain good health in commercial egg production. This paper addresses the vector capacity of the ectoparasite with special emphasis on salmonellae, pathogenic agents responsible for many of the most important outbreaks of food-borne diseases worlwide. It has been experimentally shown that *D. gallinae* could act as a biological vector of *S. enteritidis* and natural carriage of these bacteria by the mite on poultry premises has also been reported. It was also found that *D. gallinae* carried other pathogens such as *E. coli, Shigella* sp., and *Staphylococcus*, thus increasing the list of pathogenic agents potentially transmitted by the mite.

Keywords *Dermanyssus gallinae* · Vectorial role · Pathogenic agents · *Salmonella*

Introduction

It has been shown that Acari can be implicated in the vectorial transmission of diseases. However, their role in the natural transmission cycles of pathogenic agents is poorly known

C. Valiente Moro (✉)
Laboratoire « Microorganismes: Génome et Environnement », CNRS UMR, 6023 Aubiere, France
e-mail: valienteclaire@yahoo.fr

C. J. De Luna · A. Tod · J. H. Guy · O. A. E. Sparagano
School of Agriculture, Food and Rural Development,
Newcastle University, Newcastle upon Tyne NE1 7RU, UK

L. Zenner
Laboratoire de Biométrie et Biologie Evolutive, Ecole Nationale Vétérinaire de Lyon & Université de Lyon—CNRS UMR, 5558 Marcy L'Etoile, France

and the Acari, particularly mites, are largely ignored as vectors of human or animal diseases.

Within the order Acari, the Dermanyssoidea superfamily represents a vast group of ubiquitous organisms, they also have a very broad range of hosts and they can easily parasitise other species including farm animals and man. Among the superfamily, *Dermanyssus gallinae* (the poultry red mite) and *Ornithonyssus bacoti* are the only two mites for which the vectorial transmission of pathogens seems to be highly complex, i.e. not limited simply to a mechanical role of a vector that simply carries a microorganism without replication occurring (Valiente Moro et al. 2005). There are few reported studies which deal with the vectorial role of *D. gallinae* and the studies which have been published are often incomplete which could lead to an underestimation of the importance of this agent in the transmission of bacteria and viruses, responsible for animal infections and zoonoses (Table 1). For some of them, only the isolation of pathogens from field samples has been reported. However, the simple detection of a vector-borne bacterial agent in an ectoparasite does not demonstrate vector competence. It is the case for the bacteria *Erysipelothrix rhusiopathiae*, *Salmonella gallinarum* and *Listeria monocytogenes* and the Newcastle disease virus (NDV); (Grebenyuk et al. 1972; Zeman et al. 1982; Arzey 1990; Chirico et al. 2003; Valiente Moro et al. 2009). For other pathogens which have been associated with *D. gallinae*, only experimental transmissions under laboratory conditions have been carried out. Natural carriage of the pathogens by the mite was not investigated by the authors.

Table 1 Bacteria and viruses likely to be associated with *D. gallinae*

	Pathogens	Isolation from mites	Experimental transmission not demonstrated	Experimental transmission demonstrated	Related references
Virus	Avian Paramyxovirus type 1 Newcastle disease	√			Arzey (1990)
	Saint-Louis Encephalitis Virus (Flavivirus)		√		Chamberlain et al. (1957)
	Tick-Borne encephalitis Virus (Flavivirus)		√		Wegner (1976)
	Fowl Poxvirus Smallpox			√	Shirinov et al. (1972)
	Eastern Equine Encephalitis Virus (Togavirus)			√	Durden et al. (1993)
	Western Equine Encephalitis Virus (Togavirus)			√	Chamberlain and Sikes (1955)
	Venezualan Equine Encephalitis Virus (Togavirus)			√	Durden et al. (1992)
Bacteria	*Pasteurella multocida*			√	Petrov (1975)
	Erysopelothrix rhusiopathiae	√			Chirico et al. (2003)
	Salmonella gallinarum	√			Zeman et al. (1982)
	Listeria monocytogenes	√			Grebenyuk et al. (1972)
	Coxiella burnetii			√	Zemskaya and Pchelkina (1967)
	Spirochetes			√	Ciolca et al. (1968)

Some assays remained fruitless as *D. gallinae* was unable to be infected or to transmit the pathogens, supporting the relative unimportance of this mite in the transmission of active pathogen infection. This is the case for viruses responsible for tick-borne encephalitis and Saint-Louis encephalitis (Chamberlain et al. 1957; Zemskaya and Pchelkina 1962; Wegner 1976). Concerning *Pasteurella multocida*, the bacterium causing pasteurellosis, microbiological studies and biological experiments revealed that bacteria persisted in the body of *D. gallinae* mites after they engorged on infected birds (Petrov 1975). For *Coxiella burnetii*, a bacterium responsible for Q fever, Zemskaya and Pchelkina (1967) showed that *D. gallinae* could acquire infection, while feeding on infected animals. The rickettsiae survived in the mites, which subsequently fed on healthy birds, for about 6 months, and for about 1 year in dead mites. The role of the poultry red mite was also studied in the transmission of spirochaetosis, a disease caused by the bacterium *Borrelia anserine* which can infect chickens, turkeys, geese, ducks, pheasants, grouse and canaries with morbidity and mortality up to 100%. It is usually transmitted by *Argas persicus* ticks and occasionally by infected faeces. Ciolca et al. (1968) observed that spirochetes were regularly transmitted to healthy hens, provided that the mites fed on the healthy hens within 48 h after the mites had become infected. The spirochaetes were usually eliminated in the excreta shortly after ingestion suggesting that the mite was only an occasional vector of them. With the species *Spirochaeta gallinarum*, Reshetnikov (1967) observed similar results except that the interval between blood meals should not exceed 48 h in order to reproduce the disease in the host. Several laboratory studies have also concerned experimental transmissions with equine encephalitis viruses. Chamberlain and Sikes (1955) and Durden et al. (1993) showed that *D. gallinae* which engorged on chicks infected with east equine encephalitis (EEE) virus remained carriers for at least a month. Given the chronology of mean viral titres in the mite samples and the prolonged persistence of virus in the mites (30d), some viral replication may have occurred at a low level. Moreover, authors showed that mites were able to transmit the virus to other chicks by bite when taking a blood meal. Cockburn et al. (1957) obtained an infestation from *D. gallinae* which had fed on chickens infected with west equine encephalitis (WEE), but were not able to demonstrate either transmission to healthy birds or transovarian transmission in the acarian. Interesting results were also obtained by Shirinov et al. (1972), where samples of *D. gallinae* collected from poultry farms known to have birds infected with fowl pox virus were also found to harbour the virus. When naturally-infected mites were kept in the laboratory, the virus survived inside them up to 300 days. Transovarian transmission was demonstrated and the disease was transmitted to healthy fowl by the bite of infected mites.

As a result, the role of the poultry red mite as a potential mechanical vector has been clearly shown for some pathogens even if its precise role in the epidemiology of the associated pathology remains to be determined. Most of the studies are incomplete for conclusions to be made about the precise role of *D. gallinae* in the circulation of pathogens, as complete transmission ways remain to be proven as shown in Fig. 1. To compensate for this lack, a complete study was recently undertaken, exploring in detail the role of *D. gallinae* as a vector of *Salmonella*, responsible for the most often encountered zoonotic diseases in man. In 2004, 192,703 salmonellosis cases were reported in the European Union corresponding to an increase of 22% compared to 2003 and the most important increase in incidence since 1999 (European Food Safety Autority 2005). Poultry products are among the most important sources of *Salmonella* that can be transmitted through the food chain to humans (Lacey 1993). This particular serovar, strongly associated with the production of eggs for human consumption, has replaced *S. typhimurium* as the primary cause of salmonellosis worlwide (Baumler et al. 2000). Consequently, many food safety

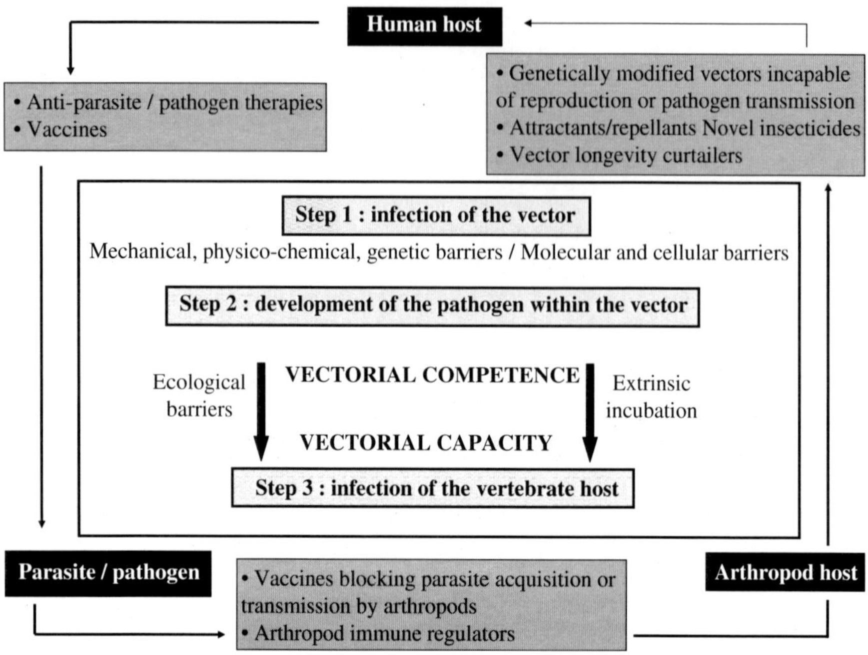

Fig. 1 Schematic representation showing the typical transmission cycle of a vector-borne parasite or pathogen between a human host and an arthropod vector, and potential steps for intervention. Different steps of the vectorial transmission are shown in the case of a biological vector

laws and hygiene control methods are aimed at preventing its transmission (Rodrigue et al. 1990). A recent hypothesis suggests that the eradication of *S. gallinarum*, a bacterium which does not affect man, would be the origin of the implantation of *S. enteritidis* in fowls which can infect man (Velge et al. 2005). These Enterobacteriaceae take part from normal flore within the chicken and, although not necessarily harmful for them, they are responsible for some of the most widespread zoonoses in the world. They are particularly resistant in the environment, probably due to their capacity to survive dessication better than other coliforms (Morse and Duncan 1974). Salmonellas, and in particular the serotype Enteritidis, survive in various wildlife reservoirs, and their presence in arthropods as litter beetles, houseflies and cockroaches has been recorded and found to be of significant importance in their transmission (Olsen and Hammack 2000; Davies and Breslin 2003; Fischer et al. 2003; Skov et al. 2004). Zeman et al. (1982) have already shown that *D. gallinae* could shelter salmonellas (serotype Gallinarum) for more than 4 months. However, authors did not investigate further by evaluating vectorial competence from the red mite for *Salmonella*. Consequently, these preliminary results raised the question of the vectorial role of *D. gallinae* for salmonella. Indeed, *D. gallinae* often hide under the dry droppings of the hens which are also frequently contaminated by the salmonellas (Morse and Duncan 1974). Moreover, mites can feed several times in each life stage on birds which increases the chance of exposure to blood-borne bacterial agents and it has been observed that the birds ingest mites (Kilpinen 2005). So the recurring problems of salmonellosis in housing systems for laying hens associated with the simultaneous presence of *D. gallinae* in these buildings, lead to a preliminary study of the role of *D. gallinae* in the transmission of salmonella, particularly *S. enteritidis*, serotype usually found in collective alimentary toxi-infections (Valiente Moro et al. 2007a, b).

Due to their haematophagous behaviour and ability to fast for long periods of time, these mites are particularly well adapted candidates for pathogen transmission. Moreover, the current lack of effective measures to control this mite, partly due to increasing acaricide resistance, causes reoccurring mite problems in poultry facilities once they have become infested by this parasite. Consequently, mites can persist in the poultry house between flock cycles and may also act as reservoir hosts for pathogenic agents such as bacteria giving a source of infection for the replacement pullets. As a result, the ubiquitous presence of *D. gallinae* in poultry breeding farms worldwide raises the question of their role in the colonisation, survival and propagation of pathogenic agents.

Materials and methods

Dermanyssus gallinae as an experimental vector of *Salmonella enteritidis*

Dermanyssus gallinae populations were collected from laying hen breeding facilities known to be free of *Salmonella* infection. Two methods of infecting the mites were tested: infection via the blood meal and via cuticular contact. The methodology of Bruneau et al. (2001) was modified slightly to create an in vitro feeding device to infect the mites during the blood meal. To ensure that any bacteria subsequently detected were those located inside the mite, we cleaned them following the protocol described by Zeman et al. (1982) using 4% w/v paraformaldehyde followed by rinsing with sterile distilled water. A total of 50 mites were fed on blood containing 10^8 CFU/ml during 4 h feeding periods. Cuticular infection of the mites was achieved by leaving them on a dry *Salmonella* coating for 48 h at room temperature (by putting a Whatman paper in contact with a saturated salmonella culture on a petri plate, then the mite was left on the contaminated paper). To test the presence of *Salmonella* within the mites immediately after the infection, 100 and 40 mites infected, respectively by the oral route and cuticular contact were individually washed as described above and analysed by microbiological culture. To test whether *Salmonella* survived and multiplied within *D. gallinae*, bacteria were counted overtime by analysing mites at 1, 3, 7 and 14 days after infection. Concerning the infection through the blood meal, we considered that all mites with a *Salmonella* count five times higher (statistically determined) than the theoretical value of 20,000 bacteria proved bacterial multiplication. This latter value was determined by considering that the number of enterobacteriaceae inside freshly engorged mites was approximately 2×10^4 CFU, since a mite absorbs about 0.2 μl of blood, containing in our experiment a bacterial level of 10^8 CFU/ml. Similarly, to have an idea of the threshold value in mites infected after cuticular contact, the average population of salmonella inside the mites 1 day after the infection was estimated as equal to 7.6×10^3 CFU/ml. All mites with a bacterial count five times higher than this threshold value i.e. 3.8×10^4 CFU/ml proved bacterial multiplication. The effect of *Salmonella* on mite oviposition, transovarial and transstadial passages was only tested for those mites, which acquired *Salmonella* during the blood meal because it would have been necessary to wash the cuticule of contaminated mites to avoid external contaminations. In this case, the mites would have been killed during the washing process. To test whether the presence of the pathogen inside engorged females reduced the number of laying females as well as their fertility rate, comparisons of the number of eggs laid were made using 165 females engorged on either infected or uninfected blood. Assessing how *D. gallinae* could contaminate birds has been investigated in two different ways. The possibility that *D. gallinae* could contaminate the blood during a blood meal after acquiring the bacteria during a previous blood meal or contact with *Salmonella*, was first tested using the in vitro feeding device. As the ingestion of *D. gallinae* by birds has been frequently reported, the assumption that hens become contaminated with *S. enteritidis* after ingesting infected mites was also assessed (Valiente Moro et al. 2007b). In this aim, 98 1-day-old chicks were inoculated orally by 10 mites: 34 chicks received mites contaminated during the blood meal, 34 received mites contaminated by the cuticular route, and 30 received uncontaminated mites. Following oral administration of contaminated mites to chicks, the ability of *S. enteritidis* to colonise the digestive tract and to invade the internal organs of the chicks was investigated. Using direct plating, *S. enteritidis* was counted from the spleen and the liver of all birds.

Natural carriage of *Salmonella* by *Dermanyssus gallinae* in poultry farms

The second step was to assess the natural carriage of *Salmonella* by *D. gallinae* in poultry farms (Valiente Moro et al. 2007d). Preliminary field studies have been undertaken using a molecular detection tool associating a simple filter-based DNA preparation with a specific 16S rDNA *Salmonella* sp. polymerase chain reaction (PCR) amplification (Desloire et al. 2006; Valiente Moro et al. 2007c). The presence of *Salmonella* was tested in *D. gallinae* collected in two types of laying farms: six farms were currently declared positive for *Salmonella* by the French DSV (Direction des Services Vétérinaires) while 10 others had been previously declared positive. For each farm, 6–20 pools of 15 mites (a total of 249 pools) were analysed for the presence of *S. enteritidis* as described above.

Dermanyssus gallinae as the carrier of other bacteria in poultry farms

Five samples of *D. gallinae* were collected from one farm in the north-east of UK, the surface sterilised before the mites were crushed as described in De Luna et al. (2008). Serial dilutions were performed up to 10^{-3} and the dilutions were put directly into culture on several non-specific media such as LB agar, blood sheep agar, and brain heart infusion agar. Specific media for enterobacteria were also used (BPL agar and XLT-4 agar). Colonies were harvested directly and DNA was extracted using a tissue protocol of a commercial kit following the manufacturer's instructions (Qiagen, UK). This was used as the template DNA for the subsequent PCR reactions. Partial amplification of the 16S rRNA gene was undertaken using a universal primer pair: 27-F′-AGAGTTTGATCMTGGCT CAG-3′, and 1513-R 5′-AGGGYTACCTTGTTACGACTT-3′ (Weisburg et al. 1991). PCR products (of around 1,200 pb) obtained from specific PCR were sequenced by dye-labelling using the BigdyeTM dideoxy technique and either the forward or reverse primers of the original PCR reactions were used. Sequences were run on an automated DNA sequence system at the IRES Genomics service at Newcastle University, UK. The BLASTN (Basic Local Alignment Search Tool) search option of the National Centre for Biotechnology Information (NCBI) internet site (http://www.ncbi.nlm.nih.gov) was used to identify close evolutionary relatives in the GenBank database. When the similarity percent between our sequences and previously described sequences exceeded 97%, the sequences were considered as corresponding to the same species as that in the GenBank reference (Stackebrandt and Goebel 1994).

Results and discussion

The results showed that immediately after the experimental infection, *Salmonella* was found in 29% of mites infected by a bloodmeal and in 55% of mites infected by cuticular contact. Given the efficiency of paraformaldehyde in eliminating all external contaminations (Zeman et al. 1982), two hypotheses can be proposed for cuticular infection: either transcuticular passage of the bacteria and/or entry of the microorganisms through the stigmata. By either infection route, a significant increase in the number of mites carrying *Salmonella* was observed in comparison to day 0 (Fig. 2a). The fact that the number of infected mites was greater 3 or 7 days after infection than the day after infection suggests that the bacteria may multiply inside the mites. Regarding the oral route, bacterial multiplication was shown for 21 and 24% of mites at D7 and D14, respectively (Fig. 2b). For the other mites, the population of bacterial was either stable or decreased over time. Similarly, multiplication was demonstrated for mites infected by cuticular contact for 42 and 25% of mites, respectively, at D7 and D14 (Fig. 2c). Cases of decreasing number of bacteria may result from an antibacterial response of mites similar to that observed for ticks, or even from the destruction of bacteria by the digestive system of the mite (Weyer 1975; Nakajima et al. 2003). Effect of *Salmonella* on mite oviposition was also demonstrated as the number of ovipositing females was significantly lower when mites were engorged on contaminated blood than when they were fed on uncontaminated blood (31 vs. 68%, respectively, $P < 0.05$). Moreover, the fecundity rate was slightly lower for mites

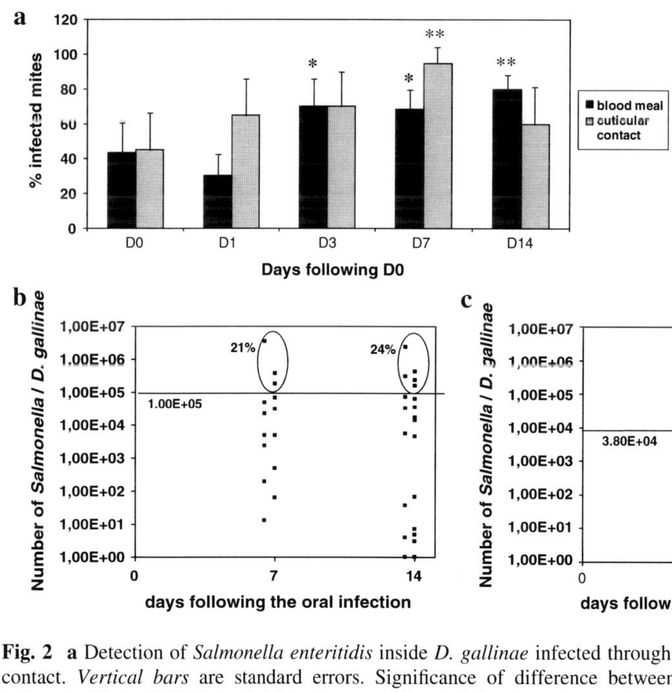

Fig. 2 **a** Detection of *Salmonella enteritidis* inside *D. gallinae* infected through a blood meal or cuticular contact. *Vertical bars* are standard errors. Significance of difference between D0 and others dates is indicated (**$P < 0.01$, *$P < 0.05$). **b** *S. enteritidis* multiplication in infected *D. gallinae* after a blood meal (theoretical value = 2.10^4 bacteria in freshly engorged mites and multiplication if the number of bacteria on SM ID = 5×2.10^4 equal to 1×10^5). **c** *S. entetitidis* multiplication in infected *D. gallinae* after cuticular contact (theoretical value = average at D1 = 7.6×10^3 and multiplication if the number of bacteria on SM ID = $5 \times 7.6.10^3$ equal to 3.8×10^4)

engorged on contaminated blood (1.31 vs. 1.92 for uninfected blood) but this difference was not significant. This decrease in fertility of *D. gallinae* could be explained by the presence of enterobacteriaceae in the reproductive organs of the mites. Of the 74 ovipositing females, 37 showed transovarial passage: in vitro females engorged on contaminated blood produced infected protonymphs. The transovarial passage has often been reported for ticks (Macaluso et al. 2001; Rennie et al. 2001), although it is less frequent in other mites. For example, *Liponyssoides sanguineus*, the main vector of *Rickettsia akari*, the bacteria responsible for vesicular fever, can transmit the bacteria to its progeny. Furthermore, out of a total of 22 N1 nymphs obtained from uncontaminated females and subsequently fed on contaminated blood, three deutonymphs were detected as positive for *Salmonella* demonstrating transstadial passage. Although passage between the deutonymph and adult stages was not tested, we can assume that it does occur, allowing bacteria to persist throughout the entire life cycle of the mite. Concerning *Salmonella* retransmission to birds, *D. gallinae* was able to contaminate the blood during a blood meal after acquiring the bacteria during a previous blood meal or contact with *Salmonella*. In cases of oral acquisition, only one *Salmonella* transmission was observed among the 18 separate assays performed. In cases of cuticular contact, blood was infected in 5 cases out of a total of 12 separate experiments. The blood on which uninfected mites were fed remained negative in each experiment. Even if a few cases of salmonellosis were observed, it may be noteworthy in infested poultry facilities due to the large number of *D. gallinae* which are often present in commercial housing conditions (Nordenfors and Chirico 2001). Concerning oral transmission of contaminated mites to chicks, faecal samples from both sets of infected chicks were positive for *Salmonella* at 6 days after inoculation while the control corresponding to chick inoculated with uncontaminated mites remained negative. On D 12 post-inoculation, *S. enteritidis* was isolated from the caecum of all birds that had received contaminated mites with an average number of *S. enteritidis* of above 8.5×10^4 MNP *Salmonella*/g (Table 2). Statistical analysis did not show any significant differences between the infection routes. The level of infection obtained in both the infection models tested shows that previously infected mites could represent a source of *Salmonella* sp. infection when eaten by 1-day-old chicks. The invasion of organs such as the liver and the spleen is an indication of systemic infection. So the reproductive organs could be also contaminated and thus it could introduce a risk for humans when eggs are consumed. Presently, it is difficult to quantify how many *D. gallinae* are ingested by hens each day

Table 2 Isolation and enumeration of *Salmonella enteritidis* in spleen, liver and caecum on day 12 post-inoculation

	Spleen		Liver		Caecum[c]	
	(CFU/g tissue) (×5)		(CFU/g tissue) (×5)		(CFU/g tissue) (×1)	
	% Positive[a]	CFU/g ± S·D[b]	% Positive[a]	CFU/g ± S·D[b]	% Positive[a]	MPN *Salmonella*/g[b]
Engorged mites	100	$2.7 \pm 1.8 \times 10^3$	100	$1.9 \pm 1.9 \times 10^2$	100	$>8.6 \times 10^4$
Cuticular mites	100	$3.6 \pm 3.5 \times 10^3$	100	$1.8 \pm 2.9 \times 10^3$	100	$>8.6 \times 10^4$

[a] Percentage of positive birds

[b] With the number of CFU per gram of tissue

[c] Caecal walls and contents estimated using MPN method

and therefore it is difficult to assess the real impact of *D. gallinae* in laying hen system facilities. However, it has been shown that this mode of infection is possible in 1-day-old chicks, even if it would be interesting to confirm whether such infections takes place in older birds. Although *D. gallinae* can both acquire *S. enteritidis* either by contaminating the blood under experimental conditions or chicks by oral ingestion, this does not mean that the mite is a natural vector. To rule out the possibility of this type of contamination in the field, it was necessary to study the level of *Salmonella* sp. in poultry facilities both before and after the arrival of a new flock. Of the six currently infected farms, only one farm had four positive pools (out of six) for *Salmonella* sp. Of the 10 farms that had previously been declared infected, only one farm had one positive pool (out of six). Thus, despite limited sampling, we have demonstrated that *D. gallinae* can be naturally infected in a contaminated poultry environment. Interestingly, *Salmonella* was detected in mites collected from a farm that was not currently contaminated, which suggests that mites infected during a previous outbreak survived the sanitation periods as well as the cleaning and disinfection programmes and may be a source of infection for replacement birds. *D. gallinae* could therefore act as a potential reservoir and will certainly play a role in the epidemiology of avian salmonellosis, as previously suggested by Zeman et al. (1982). This preliminary field study shows that *D. gallinae* can naturally harbour the pathogen and allows its persistence between outbreaks. In order to accurately evaluate the prevalence of *D. gallinae* carrying *Salmonella* in poultry farms on a national scale, much more widespread testing of facilities would be required. Other data would need to be considered in order to understand the role of *D. gallinae* in the epidemiology of avian salmonellosis, such as the vectorial capacity or the extrinsic incubation. In an epidemiological context, it is possible that infected mites crushed or eaten by chicks may be the main source of infection rather than the mite blood meal. Carriage of *Salmonella* by arthropods has already been reported in previous studies. Most recorded examples refer to litter beetles (*Alphitobius diaperinus*) or cockroaches but their role in the transmission of infection remains unproven (Davies and Wray 1955; McAllister et al. 1994; Gray et al. 1999). Ash and Greenberg (1980) showed that the German cockroach, *Blatella germanica*, is an effective mechanical transmitter of *S. typhimurium* via faeces, although the bacteria were recoverable from its gut for about 10 days longer than from the faeces.

Finally it was also found that *D. gallinae* was carrying *E. coli*, *Streptomyces* sp. and *Staphyloccocus* sp. (Table 3), the latter has been associated with infections in starlings (Berger et al. 2003). *E. coli* has been associated with Avian Pathogenic *E. coli* and Uropathogenic *E. coli*. However, there was no conclusive evidence of these organisms acting as disease causing at the poultry farm at the moment the mites were sampled.

Conclusions

In view of these results, *D. gallinae* certainly plays an important role in the transmission of salmonellosis and other diseases in poultry farms, particularly between successive flocks. It would be of further interest to quantify the level of competence of *D. gallinae* for *Salmonella* and other bacterial genera and to assess the relationships between pathogen/vector. Another perspective would be to analyse the development cycle of the pathogen within the mite by studying the crossing of intestinal and salivary barriers. The nature of the immune response of birds and its impact on the vectorial competence of mites would also be interesting to aspects to investigate. The literature reviewed and the new data presented in this paper suggest that effective control of *D. gallinae* during downtime sanitation periods

Table 3 BLAST queries showing selected closest similarities for the bacteria isolated from the different media

Media from which bacteria were isolated	Accession number	Organism	Maximum identity (%)
Brain Heart infusion	EU289096.1	Uncultured *Escherichia* sp. clone 8817-D4-C-10B	99
	CP000468.1	*Escherichia coli* APEC O1, complete genome	99
	CP000243.1	*Escherichia coli* UTI89, complete genome	99
	AE014075.1	*Escherichia coli* CFT073, complete genome	99
	AB269763.1	*Escherichia coli* gene for 16S rRNA, strain:AE1-2	99
Green Bromophenol (BPL)	EF560786.1	*Escherichia coli* strain BE45	99
	CP000468.1	*Escherichia coli* APEC O1, complete genome	99
	DQ683069.1	*Escherichia coli* strain ATCC 25922	99
	CP000247.1	*Escherichia coli* 536, complete genome	99
	CP000243.1	*Escherichia coli* UTI89, complete genome	99
XLT-4	EU289096.1	Uncultured *Escherichia* sp. clone 8817-D4-C-10B	99
	EF560775.1	*Escherichia coli* strain 246	99
	CP000468.1	*Escherichia coli* APEC O1, complete genome	99
	CP000247.1	*Escherichia coli* 536, complete genome	99
	CP000243.1	*Escherichia coli* UTI89, complete genome	99
Blood Sheep agar	EU095643.1	*Staphylococcus xylosus* strain NT-W	100
	AP008934.1	*Staphylococcus saprophyticus* subsp. *saprophyticus* ATCC 15305	99
	AF515587.1	*Staphylococcus xylosus*	99
	AB166961.1	*Staphylococcus* sp. NT N1	99
	EU430992.1	*Staphylococcus saprophyticus* isolate 23	99
	EU095643.1	*Staphylococcus xylosus* strain NT-W	100
LB agar	AB184106.1	*Streptomyces brasiliensis*	99
	EU569313.1	*Staphylococcus* sp. LV 15	99
	EU266748.1	*Staphylococcus xylosus* strain NY-5	99
	AM882700.1	*Staphylococcus xylosus*	99
	AP008934.1	*Staphylococcus saprophyticus* subsp. *saprophyticus* ATCC 15305	99

and before the introduction of new birds are key areas to target to reduce the persistence of pathogens such as *Salmonella* between successive poultry flocks.

Acknowledgments This work was partly supported financially by the European Commission through the STREP project "RESCAPE", contract no. 036018, under the 6th Framework Programme, priority 5, food quality and safety.

References

Arzey G (1990) Mechanism of spread of Newcastle disease. Technical bulletin—New South Wales, Agriculture and Fisheries 42, Sydney, 12 pp

Ash N, Greenberg B (1980) Vector potential of the German cockroach (Dictyoptera: Blatellidae) in dissemination of *Salmonella enteritidis* serotype Typhimurium. J Med Entomol 17:417–423
Baumler AJ, Hargis BM, Tsolis RM (2000) Tracing the origins of *Salmonella* outbreaks. Science 287:50–52. doi:10.1126/science.287.5450.50
Berger S, Disko R, Gwinner H (2003) Bacteria in starling nests. J Fur Ornit 144:317–322
Bruneau A, Dernburg A, Chauve C, Zenner L (2001) First in vitro cycle of the chicken mite, *Dermanyssus gallinae* (DeGeer 1778), utilizing an artificial feeding device. Parasitology 123:583–589. doi: 10.1017/S0031182001008836
Chamberlain RW, Sikes RK (1955) Laboratory investigations of the role of bird mites in the transmission of eastern and western equine encephalitis. Am J Trop Med Hyg 4:106–118
Chamberlain RW, Sikes RK, Sudia RW (1957) Attempted laboratory infection of bird mites with the virus of Saint Louis encephalitis. Am J Trop Med Hyg 6:1047–1053
Chirico J, Eriksson H, Fossum O, Jansson D (2003) The poultry red mite, *Dermanyssus gallinae*, a potential vector of *Erysipelothrix rhusiopathiae* causing erysipelas in hens. Med Vet Entomol 17:232–234. doi: 10.1046/j.1365-2915.2003.00428.x
Ciolca AL, Tanase I, May I (1968) Role of the poultry red mite, *Dermanyssus gallinae*, in the transmission of spirochaetosis. Arch Vet Pol 5:207–215
Cockburn TA, Sooter CA, Langmuir AD (1957) Ecology of western Equine and Saint Louis Encephalitis viruses. A summary of field investigations in weld county, Colorado 1949–1953. Am J Hyg 65:130–146
Davies RH, Breslin M (2003) Persistence of *Salmonella enteritidis* phage type 4 in the environment and arthropod vectors on an empty free-range chicken farm. Environ Microbiol 5:79–84. doi: 10.1046/j.1462-2920.2003.00387.x
Davies RH, Wray C (1955) The role of the lesser mealworm beetle (*Alphitobius diaperinus*) in carriage of *Salmonella enteritidis*. Vet Rec 137:407–408
De Luna CJ, Valiente Moro C, Guy JH, Zenner L, Sparagano OAE (2008) Endosymbiotic bacteria living inside the poultry red mite (*Dermanyssus gallinae*): a possible tool for biological control? Exp Appl Acarol. doi:10.1007/s10493-008-9230-2
Desloire S, Valiente Moro C, Chauve C, Zenner L (2006) Comparison of four methods of extracting DNA from *D. gallinae* (Acari: Dermanyssidae). Vet Res 37:725–732. doi:10.1051/vetres:2006031
Durden LA, Linthicum KJ, Turell MJ (1992) Mechanical transmission of Venezuelan equine encephalomyelitis by hematophagous mites (Acari). J Med Vet Entomol 9:118–121
Durden LA, Linthicum KJ, Monath P (1993) Laboratory transmission of eastern equine encephalomyelitis virus to chickens by chicken mites (Acari: Dermanyssidae). J Med Entomol 30:281–285
European Food Safety Autority (2005) The community summary report on trends and sources of zoonoses, zoonotic agents and antimicrobial resistance in the European union in 2004. 25 December 2005. Part 3.1. Salmonella, pp 1–67
Fischer OA, Matlova L, Dvorska L, Svastova P, Pavlik I (2003) Nymphs of the oriental cockroach (*Blatta orientalis*) as passive vectors of causal agents of avian tuberculosis and paratuberculosis. Med Vet Entomol 17:145–150. doi:10.1046/j.1365-2915.2003.00417.x
Gray JP, Maddox CW, Tobin PC, Gummo JD, Pitts CW (1999) Reservoir competence of *Carcinops pumilio* for *Salmonella enteritidis*. J Med Entomol 36:888–891
Grebenyuk RV, Chirov PA, Kadysheva AM (1972) The role of wild animals and blood-sucking arthropods in the epizootiology of infection with Listeria. Rol' Dikikh Zhivotnykh i Krovososushchikh Chlenistonogikh v Epizootologii Listerioza. Frunze, Kirghiz SSR; Izdatel'stvo Ilim. Institut Biologii, Akademiya Nauk Kirgizskoi SSR, Frunze, Kirghiz, SSR. 124 pp
Kilpinen O (2005) How to obtain a bloodmeal without being eaten by a host: the case of poultry red mite, *Dermanyssus gallinae*. Physiol Entomol 30:232–240. doi:10.1111/j.1365-3032.2005.00452.x
Lacey RW (1993) Food-borne bacterial infections. Parasitology 107:S75–S93
Macaluso KR, Sonenshine DE, Ceraul SM, Azad AF (2001) Infection and transovarial transmission of rickettsiae in *Dermacentor variabilis* ticks acquired by artificial feeding. Vector-Borne Zoonot 1:45–53. doi:10.1089/153036601750137660
McAllister JC, Steelman CD, Skeeles JK (1994) Reservoir competence of the lesser mealworm (Coleoptera: Tenebrionidae) for *Salmonella typhimurium* (Eubacteriales: Enterobacteriaceae). J Med Entomol 31:369–372
Morse EV, Duncan MA (1974) Salmonellosis-an environmental health problem. JAMA 175:1015–1019
Nakajima Y, Ishibashi J, Yukuhiro F, Asaoka A, Taylor D, Yumakawa M (2003) Antibacterial activity and mechanism of action of tick defensin against Gram-positive bacteria. Biochim Biophys Acta 1624:125–130

Nordenfors H, Chirico J (2001) Evaluation of a sampling trap for *Dermanyssus gallinae* (Acari: Dermanyssidae). J Econ Entomol 94:1617–1621

Olsen AR, Hammack T (2000) Isolation of *Salmonella* sp. from the housefly, *Musca domestica*, and the dumpfly (*Hydrotaea aenescens*) (Wiedmann) (Diptera:Muscidae) at caged-layer houses. J Food Prot 63:958–960

Petrov D (1975) Study of *Dermanyssus gallinae* as a carrier of *Pasteurella multocida*. Veterinarno—Mededitsinski Nauki 12:32–36

Rennie L, Wilkinson PJ, Mellor PS (2001) Transovarial transmission of African swine fever virus in the argasid tick *Ornithodoros moubata*. Med Vet Entomol 15:140–146. doi:10.1046/j.1365-2915.2001.00282.x

Reshetnikov PT (1967) The mite *Dermanyssus gallinae*, a vector of fowl spirochaetosis. Veterinariia 44:48

Rodrigue DC, Tauxe RV, Rowe B (1990) International increase in *Salmonella enteritidis*: a new pandemic. Epidemiol Infect 105:21–27

Shirinov FB, Ibragimova AI, Misirov ZG (1972) The dissemination of the fowl-pox by the mite *Dermanyssus gallinae*. Veterinarya 4:48–49

Skov MN, Spencer AG, Hald B, Petersen L, Nauerby B, Carstensen B, Madsen M (2004) The role of litter beetles as potential reservoir for *Salmonella* enterica and thermophilic *Campylobacter* spp. between broiler flocks. Avian Dis 48:9–18. doi:10.1637/5698

Stackebrandt E, Goebel BM (1994) Taxonomic note: a place for DNA–DNA reassociation and 16S rRNA sequence analysis in the present species definition in bacteriology. Int J Syst Bacteriol 44:846–849

Valiente Moro C, Chauve C, Zenner L (2005) Vectorial role of some Dermanyssoid mites (Acari, Mesostigmata, Dermanyssoidea). Parasite 12:99–109

Valiente Moro C, Chauve C, Zenner L (2007a) Experimental infection of *Salmonella enteritidis* by the poultry red mite, *Dermanyssus gallinae*. Vet Parasitol 146:329–336. doi:10.1016/j.vetpar.2007.02.024

Valiente Moro C, Fravalo P, Amelot C, Chauve C, Salvat G, Zenner L (2007b) Colonization and organ invasion in chicks experimentally infected with *Dermanyssus gallinae* contaminated by *Salmonella enteritidis*. Avian Pathol 36:307–311. doi:10.1080/03079450701460484

Valiente Moro C, Desloire S, Vernozy-Rozand C, Chauve C, Zenner L (2007c) Comparison of the VIDAS system, FTA filter-based PCR and culture on SM ID for detecting *Salmonella* in *Dermanyssus gallinae*. Lett Appl Microbiol 44:431–436. doi:10.1111/j.1472-765X.2007.02119.x

Valiente Moro C, Desloire S, Chauve C, Zenner L (2007d) Detection of *Salmonella* sp. in Dermanyssus gallinae using an FTA filter-based PCR. Med Vet Entomol 21:148–152. doi:10.1111/j.1365-2915.2007.00684.x

Valiente Moro C, Normand P, Thioulouse J, Chauve C, Zenner L (2009) Bacterial taxa associated with the hematophagous mite, *Dermanyssus gallinae* detected by 16S rDNA PCR amplification and TTGE fingerprint. Res Microbiol 160:63–70. doi:10.1016/j.resmic.2008.10.006

Velge P, Cloeckaert A, Barrow P (2005) Emergence of *Salmonella* epidemics: the problems related to *Salmonella enterica* serotype Enteritidis and multiple antibiotic resistance in other major serotypes. Vet Res 36:267–288. doi:10.1051/vetres:2005005

Wegner Z (1976) Laboratory study on some parasitic hematophagous arthropods as possible subsidiary links of the biocenosis of tick-borne encephalitis. Bull Inst Marit Trop Med Gdynia 27:75–85

Weisburg WG, Barns SM, Pelletier DA, Lane DJ (1991) 16S ribosomal DNA amplification for phylogenetic study. J Bacteriol 173(2):697–703

Weyer F (1975) Beobachtungen uber das Verhalten des Q-Fieber-Erregers (*Coxiella burneti*) in der Lederzecke Ornithodoros moubata. Tropenmed Parasitol 26:219–231

Zeman P, Stika V, Skalka B, Bartik M, Dusbabek F, Lavickova M (1982) Potential role of *Dermanyssus gallinae* (De Geer, 1778) in the circulation of the agent of pullurosis-typhus in hens. Folia Parasitol (Praha) 29:371–374

Zemskaya AA, Pchelkina AA (1962) The relation of gamasid mites to the virus of tick-borne encephalitis. Medskaya Parazitology 31:439–442

Zemskaya AA, Pchelkina AA (1967) Gamasoid mites and Q fever. In: Markevich (ed) Problemy Parazitologii, Kiev, pp 258–259

Endosymbiotic bacteria living inside the poultry red mite (*Dermanyssus gallinae*)

Carlos J. De Luna · Claire Valiente Moro · Jonathan H. Guy · Lionel Zenner · Olivier A. E. Sparagano

Originally published in the journal Experimental and Applied Acarology, Volume 48, Nos 1–2, 105–113.
DOI: 10.1007/s10493-008-9230-2

Abstract This study investigated the endosymbiotic bacteria living inside the poultry red mite collected from five samples of one commercial farm from the UK and 16 farms from France using genus-specific PCR, PCR-TTGE and DNA sequencing. Endosymbiotic bacteria are intracellular obligate organisms that can cause several phenotypic and reproductive anomalies to their host and they are found widespread living inside arthropods. The farm sampled from the UK was positive for bacteria of the genera *Cardinium* sp. and *Spiroplasma* sp. From France, 7 farms were positive for *Cardinium* sp., 1 farm was positive for *Spiroplasma* sp., 1 farm was positive for *Rickettsiella* sp. and 2 farms were positive for *Schineria* sp. However, it was not possible to detect the presence of the genus *Wolbachia* sp. which has been observed in other ectoparasites. This study is the first report of the presence of endosymbionts living inside the poultry red mite. The results obtained suggest that it may be possible that these bacterial endosymbionts cause biological modifications to the poultry red mite.

Keywords Endosymbiont · *Cardinium* · *Spiroplasma* · *Rickettsiella* · *Schineria* · Poultry red mite · *Dermanyssus gallinae*

C. J. De Luna · J. H. Guy · O. A. E. Sparagano
School of Agriculture, Food and Rural Development, Newcastle University,
Newcastle Upon Tyne NE1 7 RU, UK

C. V. Moro
Laboratoire "Microorganismes: Génome et Environnement", CNRS UMR, 6023 Aubière, France

L. Zenner
Laboratoire de Biométrie et Biologie Evolutive, Ecole Nationale Vétérinaire de Lyon &
Université de Lyon, CNRS UMR, 5558 Marcy L'Etoile, France

Present Address:
C. J. De Luna (✉)
School of Equine and Animal Science, Writtle College, Lordship Campus, Writtle,
Chelmsford CM1 3RR, UK
e-mail: carlos.deluna@writtle.ac.uk

Introduction

The poultry red mite, *Dermanyssus gallinae*, is considered to be the most important haematophagous ectoparasite of laying hens in Europe (Chauve 1998). An infestation of red mite can cause a reduction in egg production, blood staining of eggs, anaemia and may even cause the death of the host (Chauve 1998), resulting in significant economic losses and compromised welfare within poultry flocks. Although the importance of the poultry red mite as a direct parasite is well documented, its role as a disease vector may also be important but it is not well understood (Chirico et al. 2003; Valiente Moro et al. 2005).

Control of red mite infestations has traditionally been achieved by treating buildings and farm infrastructures with synthetic compounds with acaricidal activity (such as permethrin, carbaryl or amitraz). However, it has been reported that the poultry red mite has developed resistance to many of these products (Beugnet et al. 1997; Marangi et al. 2008). Many synthetic acaricides have also been regarded as environmentally toxic agents that can be absorbed through the egg shell and pass into the food chain. Thus the use of a number of these products is presently being restricted in the EU and other countries (Chirico and Tauson 2002). There is a need to identify and implement novel control methods for the poultry red mite.

Arthropods have been found to be infected with endosymbiotic bacteria, and in recent years, evidence has been gathered on the importance of these bacterial populations that live inside arthropods (Sparagano and De Luna 2008). Endosymbiotic infection to arthropods in general produce different phenotypic and reproductive anomalies to the hosts, e.g., parthenogenesis (Zchori-Fein et al. 2001; Zchori-Fein and Perlman 2004), reproductive incompatibilities between infected and uninfected individuals (Hunter et al. 2003), disturbance of oogenesis (Zchori-Fein et al. 2001), causing mortality of male embryos (Werren et al. 1986), and can also give rise to populations consisting only of haploid individuals (Weeks et al. 2001). Across the literature, the genus *Wolbachia* sp., *Cardinium* sp. and *Spiroplasma* sp. have been reported to be associated with arthropods (Fujii et al. 2004; Morimoto et al. 2006; Enigl and Schausberger 2007; Gotoh et al. 2007a, b).

Cytoplasmic incompatibility (CI) has been reported to occur in insects and arthropods infected with *Wolbachia* sp. (Charlat et al. 2001; Gotoh et al. 2007b; O'Neill et al. 1992) and also with *Cardinium* sp. (Hunter et al. 2003; Gotoh et al. 2007a). In the case of the ash whitefly parasitoid *Encarsia inaron*, mixed infections with both *Wolbachia* sp. and *Cardinium* sp. also produced CI (Perlman et al. 2006). CI is a reproductive incompatibility between infected males and females that are either uninfected, or infected with different strains of the endosymbiont (Gotoh et al. 2007a, b). The symbiotic bacteria spread in an arthropod population causing a reduction in fitness through failed matings (Perlman et al. 2006). Early death of offspring has been reported when *Cardinium*-infected males mated with uninfected females (Hunter et al. 2003; Gotoh et al. 2007a).

Amongst other effects, it has been reported that *Spiroplasma* sp. causes male-mortality by producing low hatchability and mortality of male offspring during the reproduction of infected females and causing selection towards feminised hosts (Hurst et al. 1999; Pool et al. 2006; Tinsley and Majerus 2006).

Regarding *Schineria* sp. the scientific literature only reports the presence of this species of bacteria on the larval stages of the myasis-causing fly *Wohlfahrtia magnifica* (Toth et al. 2001, 2006). They have suggested that the effects on the fly are related to the change of different stages during its metamorphosis (Toth et al. 2006). *Ricketsiella* sp. has been isolated from a predatory mite species *Phytoseiulus persimilis* (Sutakova and Arutunyan 1990), however, they did not report the effects of this bacterial species upon the biology of the mite.

Biological control, using endosymbiotic bacteria-derived substances that may produce changes to the reproduction of arthropods, may be a viable alternative to traditional methods of control of the poultry red mite. This paper describes an investigation into the identification of endosymbiotic bacteria which live inside the poultry red mite.

Materials and methods

Poultry red mites

Poultry red mites were obtained from one commercial poultry farm in the UK at five different times during a productive cycle. One pool of samples per collection time was obtained (total 5 pools). From France 16 farms were sampled. From those 16 farms, 15 individuals from each farm were analysed (total 240 individuals) and 5 pools of 10 mites per farm were created (total 80 pools). Mites were collected either directly from farms in plastic bags and transported alive to the laboratory or were sent by post to the laboratory in plastic containers and maintained in the dark at approximately 22°C for a period of time (at least for 10 days) such as they were considered to be unfed, i.e., until the blood meal was fully digested.

Red mite preparation

To confirm that the infection of bacteria originated from inside the mites, the exterior surface of the mite was sterilised. The mites from the UK were prepared as follows: a sample of 200–500 unfed mites were collected and transferred to a sterile 1.5 ml microcentrifuge tube. The mites were washed by adding five volumes of phosphate buffered saline solution pH 7.0 to the microcentrifuge tube. This was repeated three times. Next the sample was washed twice in five volumes of 5% sodium hypochlorite solution. Finally, the mite sample was washed five times with ultra pure water containing Triton X-100 to remove any traces of sodium hypochlorite. The sample was then transferred to a sterile 1.5 ml micro-centrifuge tube and crushed in 50 μl of ultrapure water. Mites originated from France were disinfected as previously described by 1% (w/vol) paraformaldehyde followed by rinsing with sterile distilled water (Valiente Moro et al. 2007).

DNA extraction and PCR amplification of endosymbiotic bacteria DNA

Poultry red mite DNA was extracted following the tissue protocol of a commercial kit (Qiagen, UK) as described in Desloire et al. (2006). This was used as the template DNA for the subsequent PCR reactions except where indicated below. DNA from French mites were amplified by PCR following by separation with temporal temperature gel electrophoresis (TTGE) while DNA samples from UK were amplified by genus-specific PCR reactions due to the technological capacity in each collaborating group.

Universal PCR amplification of 16S rDNA fragments and TTGE analysis

Genomic DNA extracted from mites originating from French farms was amplified by PCR using conserved 16S bacterial rRNA gene domain primers, 350F (5′-CTCCTACGGG AGGCAGCAGT-3′) and PC535 (5′-GTATTACCGCGGCTGCTGGCA-3′), targeting the eubacterial conserved V3 region of 16S rRNA gene, as already described by Valiente Moro et al. (2008). The first 40 nucleotides of 350F comprise the GC-clamp which prevented strand dissociation at high temperature during electrophoresis.

For sequence-specific separation of PCR products, the TTGE Dcode System (BioRad, Marnes-la-Coquette, France) was used (Valiente Moro et al. 2008). Gel electrophoresis was performed for 16 h in 1x TAE running buffer, 7 M urea, with 9% acrylamide-bis-acrylamide (37.5:1) gels at a constant voltage of 60 V and with a temperature gradient from 63 to 70.5°C at a constant temperature increment of 0.4°C/h. After electrophoresis, gels were incubated using the sensitive SYBR green nucleic acid gel staining method (Roche Diagnostics, Mannheim, Germany) and DNA fragments were visualised under UV light.

Specific PCR amplification of endosymbiontic bacteria DNA

Partial DNA amplification of the 16S rRNA gene of *Cardinium* sp. was carried out by PCR using *Cardinium*-specific primers Ch-F 5′-TACTGTAAGAATAAGCACCGGC-3′, and Ch-R 5′-GTGGATCACTTAACGCTTTCG-3′ as described in Zchori-Fein and Perlman (2004). PCR cycling conditions were: initial denaturation at 95°C for 2 min, 30 cycles of denaturation at 92°C for 30 s, annealing at 55°C for 30 s and extension at 72°C for 30 s and a final extension at 72°C for 5 min. This method amplified a PCR product of around 500 bp.

Partial DNA amplification of the 16S rRNA gene of *Spiroplasma* sp. was carried out following the method of Enigl and Schausberger (2007) by using primers 27-F 5′-GAGA GTTTGATCCTGGCTCAG-3′ (described in Weisburg et al. 1991) and TKSSsp 5′-TA GCCGTGGCTTTCTGGTAA-3′ (described in Fukatsu and Nikoh 2000). PCR cycling conditions were: initial denaturation at 94°C for 2 min, 34 cycles of denaturation at 94°C for 30 s, annealing at 57°C for 30 s and extension at 72°C for 30 s and a final extension at 72°C for 5 min. This method amplified a PCR product of around 600 bp.

Finally for *Wolbachia* sp., three attempts were made to amplify DNA. Partial amplification of a homologue of the fushi-tarazu (*ftZ*) gene was made following Noda et al. (1997) by using primers *ftsZ*-F 5′-GTATGCCGATTGCAGAGCTTG-3′, and *ftsZ*-R 5′-GCCATGA GTATTCACTTGGCT-3′ (Holden et al. 1993). The second approach consisted in amplifying partial DNA from the *Wolbachia* sp. surface protein (wsp) gene following Enigl and Schausberger (2007) by using primers wsp-F 5′-TGGTCCAATAAGTGATGAAG AAACTAGCTA-3′, and wsp-R 5′-AAAAATTAAACGCTACTCCAGCTTCTGCAC-3′, described in Jeyaprakash and Hoy (2000). Finally, a hemi-nested PCR method was used as described in Weeks et al. (2003). First, partial amplification of the 16S rRNA was undertaken using a universal primer pair: 27-F 5′-AGAGTTTGATCMTGGCTCAG-3′, and 1513-R 5′-AGGGYTACCTTGTTACGACTT-3′ (Weisburg et al. 1991). Once a PCR product was obtained, partial amplification using the PCR product itself as a DNA template was attempted using *Wolbachia*-specific primers for the 16S rRNA gene: 5′-TTG TAGCCTGCTATGGTATAACT-3′, which is in the variable V1 region, and 5′-GAAT AGGTATGATTTTCATGT-3′, which is the reverse complement of the V6 region (O'Neill et al. 1992). When PCR products where obtained, these were electrophoresed in 1.0% agarose gel and stained with ethidium bromide.

DNA sequencing

Ten PCR products for each bacteria (*Cardinium* sp. and *Spiroplasma* sp.), obtained from genus-specific PCR, were sequenced by dye-labelling using the Bigdye™ dideoxy technique and the forward and reverse primers (five for each primer) of the original ten PCR reactions were used. Sequences were run on an automated DNA sequence system at the IRES Genomics service at Newcastle University, UK.

Concerning TTGE fragments, bands were excised and the DNA was eluted in 100 μl of sterile water overnight at 4°C before PCR amplification with the same eubacterial primer set except that the primer 350F was used without GC clamp as described by Valiente Moro et al. (2008). PCR products were sequenced by Genoscreen, Lille, France and then compared with known sequences listed in the GenBank nucleotide sequence databases. The BLASTN (Basic Local Alignment Search Tool) search option of the National Centre for Biotechnology Information (NCBI) internet site (http://www.ncbi.nlm.nih.gov) was used to identify close evolutionary relatives in the GenBank database. When the similarity percent between our sequences and previously described sequences exceeded 97%, the sequences were considered as corresponding to the same species as that in the GenBank reference (Stackebrandt and Goebel 1994).

Results

The farm sampled from the UK was positive for bacteria of the genera *Cardinium* sp. and *Spiroplasma* sp. Regarding the results from France, 7 farms out of 16 were positive for *Cardinium* sp., 1 out of 16 was positive for *Spiroplasma* sp., 1 out of 16 was positive for *Rickettsiella* sp. and 2 farms out of 16 were positive for *Schineria* sp. Detailed information is presented in Table 1.

All the samples from the UK showed mixed infection with *Cardinium* sp. and *Spiroplasma* sp. However, since all the samples contained at least 200 mites, it is not possible to know whether the mixed infection occurred individually, that is, if 1 single mite carried the 2 species of endosymbiont at the same time. In the case of *Wolbachia* sp., none of the three amplification methods used produced a successful amplification and thus no further attempts to screen *Wolbachia* sp. living inside the poultry red mite were made.

Nucleotide sequence accession numbers

The sequences obtained from the UK samples were deposited in the NCBI GenBank with accession numbers EU670051 for the *Cardinium* sp. sequence and EU670052 for the *Spiroplasma* sp. sequence. Only two sequences were obtained since the sequences for both the forward and the reverse primers of both organisms were 100% identical when compared between pools of samples.

Table 1 Presence of endosymbiont species living inside the poultry red mite according to number of positive farms and location

Endosymbiont species	No. of positive farms/ farms sampled	Total number of positive individuals and total number of pools	Country
Spiroplasma spp.	1/16	1 individual	France
	1/1	5 pools	UK
Cardinium spp.	7/16	34 individuals and 18 pools	France
	1/1	5 pools	UK
Ricketsiella spp.	12/16	67 individuals and 19 pools	France
Schineria spp.	2/16	4 individuals	France

Table 2 BLAST queries showing closest similarities for the *Cardinium* sp. and *Spiroplasma* sp. sequences

Symbiont/acc. no.	BLAST queries (acc. no.)	Similarity in bp (%)
Cardinium sp. (EU670051)	Uncultured *Candidatus Cardinium* sp. 16S ribosomal RNA gene, partial sequence (EU333931)	322/323 (99%)
	Endosymbiont of *Metaseiulus occidentalis* 16S ribosomal RNA gene, partial sequence (EU333931)	321/323 (99%)
	Endosymbiont of *Brevipalpus obovatus* 16S ribosomal RNA gene partial sequence (AY279401)	321/323 (99%)
	Endosymbiont of *Oligonychus ilicis* gene for 16S rRNA, partial sequence (AB241130)	321/323 (99%)
	Cardinium endosymbiont of *Tetranychus cinnabarinus* 16S ribosomal RNA gene, partial sequence (DQ449047)	319/323 (98%)
Spiroplasma sp. (EU670052)	*Spiroplasma* sp. 'Gent' 16S ribosomal RNA gene, partial sequence (AY569829)	400/400 (100%)
	Spiroplasma-symbiont of *Antonina crawii* gene for 16S rRNA, partial sequence (AB030022)	399/400 (99%)
	Uncultured *Spiroplasma* sp. clone D2.2.14 16S ribosomal RNA gene, partial sequence (AY837733)	399/400 (99%)
	Spiroplasma symbiont of *Tetranychus urticae* 16S ribosomal RNA gene, partial sequence (DQ910772)	395/400 (98%)
	Uncultured *Spiroplasma* sp. clone Hg5-23 16S ribosomal RNA gene, partial sequence (EU344933)	395/400 (98%)

BLAST queries

Table 2 shows similarity for both the *Cardinium* sp. and *Spiroplasma* sp. sequences when a BLAST query was performed for the UK samples. For both endosymbionts there was a close similarity to other endosymbionts of arthropods.

Discussion

The results demonstrated that the poultry red mite acts as a host of *Cardinium* sp. *Spiroplasma* sp., *Ricketsiella* sp. and *Shineria* sp., a finding that confirms the ubiquitous presence of reproductive endosymbiotic bacteria in arthropods. The poultry red mite is a parasitic haematophagous arthropod, and previous studies have also reported *Cardinium* sp. on predatory mites. Weeks et al. (2003) and Zchori-Fein and Perlman (2004) found the presence of the endosymbiont on *Metaseiulus occidentalis*, a mite species that predates on spider mites. *Spiroplasma* sp. has previously been identified inside the parasitic mite *Spinturnix* sp. (Reeves et al. 2006).

However, the presence of the endosymbiont *Wolbachia* sp. could not be determined. A possible explanation for encountering false negatives could be due to the concentration and the quality of the template DNA and the presence of DNA inhibitors (Jeyaprakash and Hoy 2000).

In this study a double bacterial infection on the poultry red mite was detected. The uncertainty of knowing if the mixed infection happened at the individual level has already been mentioned, it would be important to know the significance of these mixed infections. Double infection in mites has already been reported in other studies of predatory mites.

Weeks et al. (2003) and Zchori-Fein and Perlman (2004) reported mixed infections with *Wolbachia* sp. and *Cardinium* sp. (or Cytophaga-Like-Organism -CLO- as it was known back then) on the mite species *M. occidentalis*. Enigl and Schausberger (2007) also found a double infection of *Wolbachia* sp. and *Spiroplasma* sp. on the herbivorous spider mite *Tetranychus urticae*.

The poultry red mites were found to be infected with *Cardinium* sp. and *Spiroplasma* sp. Previous studies (e.g., Perlman et al. 2006) have agreed that mixed infections are relevant to the reproductive pattern of their hosts, and thus endosymbiotic bacteria may have some potential to be used as a biological control method.

Nevertheless, it is not possible to identify the significance of the presence of *Schineria* sp. *Schineria* has been isolated from the larvae of *Wohlfahrtia magnifica* (Diptera: Sarcophagidae) a myiasis-causing fly species for most domestic animals (Toth et al. 2001, 2006). Toth et al. (2006) suggested that *Schineria* larva has a strong chitinase activity and that it may contribute to the development of fly larvae and influence the metamorphosis of *W. magnifica*. However, there is no report on the scientific literature that gives information of the presence of *Schineria* sp. in arthropods; therefore to our knowledge this study is the first report that has detected the presence of *Schineria* sp. living inside an arthropod species. *Ricketsiella* sp. has been isolated from the predatory mite *P. persimilis* (Sutakova and Arutunyan 1990); however, they did not specify the effect of this bacterial species upon the biological processes of the mite.

This study is the first report on endosymbiotic bacteria living inside the poultry red mite. However, the effect of these infections on the poultry red mite is not known. Although there are examples in the literature of the non-deleterious effects of single endosymbiont infection or between crosses of infected against non-infected individuals (e.g., Montenegro et al. 2006), or that endosymbiotic infection delays the effect of viral infections to the host organism (Hedges et al. 2008) there are other studies that have successfully demonstrated the potential use of endosymbiotic infections as a control method. For example, Perlman et al. (2006) demonstrated the effect of a double infection by *Wolbachia* sp. and *Cardinium* sp. on the ash whitefly parasitoid *Encarsia inaron*. Crosses between doubly infected males and antibiotic-cured females produced cytoplasmic incompatibility by producing a decline in female progeny without a significant increase in male progeny. Recently, Perlman et al. (2008) have described the population biology of a *Cardinium* sp. infected parasitic wasp, and they found that *Cardinium* sp. is very similar to *Wolbachia* sp. in the way they produce cytoplasmic incompatibility to the host. It is therefore important to discover the significance of the endosymbiotic infections reported in the current study for the poultry red mite. Once the effects produced by these endosymbiotic bacteria to the poultry red mite have been identified and understood, there might be a possibility that these endosymbionts could be used as means of a biological control method.

Acknowledgments This work was supported financially by the European Commission through the STREP project "RESCAPE", contract no. 036018, under the sixth Framework Programme, priority 5, food quality and safety.

References

Beugnet F, Chauve C, Gauthey M, Beert L (1997) Resistance of the red poultry mite to pyrethroids in France. Vet Rec 140:577–579

Charlat S, Bourtzis K, Merçot H (2001) *Wolbachia*-induced cytoplasmic incompatibility. In: Seckbach J (ed) Symbiosis: mechanisms and model systems. Kluwer, Dordrecht, pp 621–644

Chauve C (1998) The poultry red mite *Dermanyssus gallinae* (De Geer, 1778): current situation and future prospects for control. Vet Parasitol 79:239–245. doi:10.1016/S0304-4017(98)00167-8

Chirico J, Tauson R (2002) Traps containing acaricides for the control of *Dermanyssus gallinae*. Vet Parasitol 110:109–116. doi:10.1016/S0304-4017(02)00310-2

Chirico J, Eriksson H, Fossum O, Jansson D (2003) The poultry red mite, *Dermanyssus gallinae*, a potential vector of *Erysipelothrix rhusiopathiae* causing erysipelas in hens. Med Vet Entomol 17:232–234. doi: 10.1046/j.1365-2915.2003.00428.x

Desloire S, Valiente Moro C, Chauve C, Zenner L (2006) Comparison of four methods of extracting DNA from *D. gallinae* (Acari: Dermanyssidae). Vet Res 37:725–732. doi:10.1051/vetres:2006031

Enigl M, Schausberger P (2007) Incidence of the endosymbionts *Wolbachia, Cardinium* and *Spiroplasma* in phytoseiid mites and associated prey. Exp Appl Acarol 42:75–85. doi:10.1007/s10493-007-9080-3

Fujii Y, Kubo T, Ishikawa H, Sasaki T (2004) Isolation and characterization of the bacteriophage WO from *Wolbachia* an arthropod endosymbiont. Biochem Biophys Res Commun 317:1183–1188. doi:10.1016/j.bbrc.2004.03.164

Fukatsu T, Nikoh N (2000) Endosymbiotic microbiota of the bamboo pseudococcid *Antonina crawii* (Insecta, Homoptera). Appl Environ Microbiol 66:643–650. doi:10.1128/AEM.66.2.643-650.2000

Gotoh T, Noda H, Ito S (2007a) *Cardinium* symbionts cause cytoplasmic incompatibility in spider mites. Heredity 98:13–20. doi:10.1038/sj.hdy.6800881

Gotoh T, Sugusawa J, Noda H, Kitashima Y (2007b) *Wolbachia*-induced cytoplasmic incompatibility in Japanese populations of *Tetranychus urticae* (Acari: Tetranychidae). Exp Appl Acarol 42:1–6. doi: 10.1007/s10493-007-9072-3

Hedges LM, Brownlie JC, O'Neill SL, Johnson KP (2008) *Wolbachia* and virus protection in insects. Science 322:702

Holden PR, Brookfield FY, Jones P (1993) Cloning and characterization of an *ftsZ* homologue from a bacterial symbiont of *Drosophila melanogaster*. Mol Gen Genet 240:213–220. doi:10.1007/BF00277059

Hunter MS, Perlman SJ, Kelly SE (2003) A bacterial symbiont in the *Bacteroidetes* includes cytoplasmic incompatibility in the parasitoid wasp *Encarsia pergandiella*. Proc R Soc Lond B Biol Sci 270:2185–2190. doi:10.1098/rspb.2003.2475

Hurst GDD, der Schulenburg JHG, Majerus TMO, Bertrand D, Zakharov IA, Baungaard J, Volkl W, Stothamer R, Majerus MEN (1999) Invasion of one insect species, *Adalia bipunctata*, by two different male-killing bacteria. Insect Mol Biol 8:133–139. doi:10.1046/j.1365-2583.1999.810133.x

Jeyaprakash A, Hoy MA (2000) Long PCR improves *Wolbachia* DNA amplification: wsp sequences found in 76% of sixty-three arthropod species. Insect Mol Biol 9:393–405. doi:10.1046/j.1365-2583.2000.00203.x

Marangi M, Cafiero MA, Capelli G, Camarda A, Sparagano OAE, Giangaspero A (2008) Evaluation of the poultry red mite, *Dermanyssus gallinae* (Acari: Dermanyssidae) susceptibility to some acaricides in field population from Italy. Exp Appl Acarol. doi:10.1007/s10493-008-9224-0

Montenegro H, Petherwick AS, Hurst GDD, Klaczko LB (2006) Fitness effects of *Wolbachia* and *Spiroplasma* in *Drosophila melanogaster*. Genetica 127:207–215. doi:10.1007/s10709-005-3766-4

Morimoto S, Kurtti TJ, Noda H (2006) In vitro cultivation and antibiotic susceptibility of a Cytophaga-like intracellular symbiont isolated from the tick *Ixodes scapularis*. Curr Microbiol 52:324–329. doi: 10.1007/s00284-005-0349-7

Noda H, Munderloh UG, Kurtti TJ (1997) Endosymbionts of ticks and their relationship to *Wolbachia* spp. and tick-borne pathogens of humans and animals. Appl Environ Microbiol 63:3926–3932

O'Neill SL, Giordano R, Colbert AME, Karr TL, Robertson HM (1992) 16S rRNA phylogenetic analysis of the bacterial endosymbionts associated with cytoplasmic incompatibility in insects. Proc Natl Acad Sci USA 89:2699–2702. doi:10.1073/pnas.89.7.2699

Perlman SJ, Kelly SE, Zchori-Fein E, Hunter MS (2006) Cytoplasmic incompatibility and multiple symbiont infection in the ash whitefly parasitoid, *Encarsia inaron*. Biol Control 39:474–480. doi:10.1016/j.biocontrol.2006.05.015

Perlman SJ, Kelly SE, Hunter MS (2008) Population biology of cytoplasmic incompatibility: maintenance and spread of *Cardinium* symbionts in a parasitic wasp. Genetics 178:1003–1011. doi:10.1534/genetics.107.083071

Pool JE, Wong A, Aquadro CF (2006) Finding of male-killing *Spiroplasma* infecting *Drosophila melanogaster* in Africa implies transatlantic migration of this endosymbiont. Heredity 97:27–32. doi: 10.1038/sj.hdy.6800830

Reeves WK, Dowling APG, Dasch GA (2006) Rickettsial agents from parasitic Dermanyssoidea (Acari: Mesostigmata). Exp Appl Acarol 38:181–188. doi:10.1007/s10493-006-0007-1

Sparagano OAE, De Luna CJ (2008) From population structure to genetically-engineered vectors: new ways to control vector-borne diseases? Infect Genet Evol 8:520–525. doi:10.1016/j.meegid.2007.05.002

Stackebrandt E, Goebel BM (1994) Taxonomic note: a place for DNA–DNA reassociation and 16S rRNA sequence analysis in the present species definition in bacteriology. Int J Syst Bacteriol 44:846–849

Sutakova G, Arutunyan ES (1990) The spider-mite predator *Phytoseiulus persimilis* and its association with microorganisms—an electron-microscopy study. Acta Entomol Bohemoslov 87:431–434

Tinsley MC, Majerus MEN (2006) A new male-killing parasitism: *Spiroplasma* bacteria infect the ladybird beetle *Anisosticta novemdecimpunctata* (Coleoptera: Coccinellidae). Parasitology 132:757–765. doi: 10.1017/S0031182005009789

Toth E, Kovacs G, Schumann P, Kovacs AL, Steiner U, Halbritter A, Marialigeti K (2001) *Schineria* larvae gen. nov. sp. nov., isolated from the 1st and 2nd larval stages of *Wohlfahrtia magnifica* (Diptera: Sarcophagidae). Int J Syst Evol Microbiol 51:401–407

Toth EM, Hell E, Kovacs G, Borsodi AK, Marialigeti K (2006) Bacteria isolated from the different developmental stages and larval organs of the obligate parasitic fly, *Wohlfahrtia magnifica* (Diptera: Sarcophagidae). Microb Ecol 51:13–21. doi:10.1007/s00248-005-0090-6

Valiente Moro C, Chauve C, Zenner L (2005) Vectorial role of some dermanyssoid mites (Acari, Mesostigmata, Dermanyssoidea). Parasite 12:99–109

Valiente Moro C, Fravalo P, Amelot M, Chauve C, Zenner L, Salvat G (2007) Colonization and organ invasion in chicks experimentally infected with *Dermanyssus gallinae* contaminated by *Salmonella Enteritidis*. Avian Pathol 36:307–311. doi:10.1080/03079450701460484

Valiente Moro C, Thioulouse J, Chauve C, Normand P, Zenner L (2008) Bacterial taxa associated with hematophagous mite *Dermanyssus gallinae* detected by 16S rDNA amplification and TTGE fingerprinting. Res Microbiol. doi:10.1016/j.resmic.2008.10.006

Weeks AR, Marec F, Breeuwer JAJ (2001) A mite species that consists entirely of haploid females. Science 292:2479–2482. doi:10.1126/science.1060411

Weeks AR, Velten R, Stouthamer R (2003) Incidence of a new sex-ratio-distorting endosymbiotic bacterium among arthropods. Proc R Soc Lond B Biol Sci 270:1857–1865. doi:10.1098/rspb.2003.2425

Weisburg WG, Barns SM, Pelletier DA, Lane DJ (1991) 16S ribosomal DNA amplification for phylogenetic study. J Bacteriol 173:697–703

Werren JH, Skinner SW, Huger AM (1986) Male-killing bacteria in a parasitic wasp. Science 231:990–992. doi:10.1126/science.3945814

Zchori-Fein E, Perlman SJ (2004) Distribution of the bacterial symbiont *Cardinium* in arthropods. Mol Ecol 13:2009–2016. doi:10.1111/j.1365-294X.2004.02203.x

Zchori-Fein E, Gottlieb Y, Kelly SE, Brown JK, Wilson JM, Karr TL, Hunter MS (2001) A newly discovered bacterium associated with parthenogenesis and a change in host selection behavior in parasitoid wasps. Proc Natl Acad Sci USA 98:12555–12560. doi:10.1073/pnas.221467498

Molecular phylogenetic assessment of host range in five *Dermanyssus* species

L. Roy · A. P. G. Dowling · C. M. Chauve · I. Lesna · M. W. Sabelis · T. Buronfosse

Originally published in the journal Experimental and Applied Acarology, Volume 48, Nos 1–2, 115–142.
DOI: 10.1007/s10493-008-9231-1

Abstract Given that 14 out of the 25 currently described species of *Dermanyssus* Dugès, 1834, are morphologically very close to each another, misidentifications may occur and are suspected in at least some records. One of these 14 species is the red fowl mite, *D. gallinae* (De Geer, 1778), a blood parasite of wild birds, but also a pest in the poultry industry. Using molecular phylogenetic tools we aimed to answer two questions concerning host specificity and synanthropicity: (1) is *D. gallinae* the only species infesting European layer farms?, and (2) can populations of *D. gallinae* move from wild to domestic birds and vice versa? Mitochondrial cytochrome oxidase I gene sequences were obtained from 73 *Dermanyssus* populations collected from nests of wild European birds and from poultry farms and these were analyzed using maximum parsimony and Bayesian inference. Mapping of the observed host range on the obtained topology and correlation with behavioural observations revealed that (1) host range is strongly dependent on some ecological parameters (e.g. nest hygiene, exposure to pesticides and predators), that (2) out of five species under test, synanthropic populations were found only in lineages of *D. gallinae*, and that (3) at least some haplotypes found in wild birds were very close to those found in association with domestic birds.

Keywords *Dermanyssus* · Bird parasite · mt-COI · Host range · Synanthropicity · Host transfer

---

L. Roy (✉) · C. M. Chauve · T. Buronfosse
Université de Lyon, Ecole Nationale Vétérinaire de Lyon,
Laboratoire de Parasitologie, Marcy l'Etoile, France
e-mail: l.roy@vet-lyon.fr

A. P. G. Dowling
Department of Entomology, University of Arkansas, Fayetteville, AR, USA

I. Lesna · M. W. Sabelis
University of Amsterdam, IBED, Amsterdam, The Netherlands

Introduction

Dermanyssus gallinae (De Geer, 1778) is a nidicolous mite that is well known as a pest in poultry farms. Apart from this species, there are 24 other species that are currently included in the genus *Dermanyssus* Dugès 1834 (Roy and Chauve 2007; Knee 2008; Roy et al. 2008), but 14 of these species are morphologically very similar to *D. gallinae*, known as the poultry red mite. This may lead to identification problems and molecular tools are needed to answer questions concerning host specificity and synanthropicity: is *D. gallinae* the only species infesting European farms? Can populations of *D. gallinae* from wild birds and those from domestic birds undergo genetic exchange?

Host–parasite relationships, and especially host specificity, are hard to define in many mite groups, because they are often opportunistic, and they may have multiple hosts during post-embryonic development. For example, the ticks have been divided into 6 different categories according to the relative width of their host range and according to the mono- or polyxeny during development (Hoogstraal and Aeschlimann 1982). A similar situation exists with respect to species in the genus *Dermanyssus*. Although these species seem to complete their whole life cycle on a single bird species, the amplitude of their host range varies between two groups of species. According to previous records, only a few species appeared to be specialists in a single bird family within *Dermanyssus*, such as *D. alaudae* (Schrank, 1781) (Alaudidae only), *D. quintus* Vitzthum 1921 and *D. hirsutus* Moss and Radovsky 1967 (Picidae only) (Roy and Chauve 2007). On the other hand, most of the species in the Moss' *gallinae*-group, are known for having a very large host spectrum, involving a variety of bird families, widely distributed in bird phylogeny. *D. gallinae* and *D. hirundinis* (Hermann, 1804) have been recorded in the literature, respectively, from 8 to 9 different bird orders, some of which are phylogenetically very distant. For instance, *D. gallinae* can develop in some Galliformes as well as in some Passeriformes, the former being basally and the latter distally situated in the large clade of Neognathae according to the phylogenetic reconstruction of birds proposed by Livezey and Zusi (2007). Not to mention several mammalian recorded parasitized species. However, interpretation of data available to date is blurred by misidentifications resulting from confusion on morphological discrimination and host specificity is likely to slightly differ in some species in the *gallinae*-group (Roy et al. 2008).

In France, 5 species are commonly found in wild avifauna, which all belong to Moss' *gallinae*-group (Roy et al. 2008). *D. gallinae*, the Chicken Red Mite, seems to be the only species encountered in farms and breeding facilities. Present in more than 80% in layer farms in Europe, it is an important pest, inducing sanitary problems and financial losses.

As *D. gallinae* is a haematophagous mite, a pest in poultry industry and potentially a vector of some pathogens (Valiente Moro et al. 2005, 2007), it is of practical interest to assess if there are genetic exchanges between populations hosted by wild avifauna and domestic fowl. The aim of the work described in this article is to assess host specificity and host range in five *Dermanyssus* species by using molecular tools of phylogenetic analysis. An inventory of *Dermanyssus* species collected from wild and domestic birds in France and the Netherlands is provided as a first step towards a more comprehensive analysis. Moreover, the exploration of host specificity using phylogenetic tools will be complemented by some bioassays.

Materials and methods

Field sampling

For birds that re-use the nest, part of the nests was collected after the birds had left the nest. Otherwise whole nests were collected. Most of wild birds' nests were collected in France, and a few were collected in The Netherlands and in the USA. In addition, hundreds of wild birds have been directly examined in France. Moreover, some mite populations were collected from layer hen farms (from France, Norway, Denmark, Poland, Belgium) and facilities for breeding canaries, pigeons and chickens (France, Italy, Spain). A list of DNA-tested mite populations is provided in the Appendix. A population corresponds to a group of *Dermanyssus* mites found from a single nest (or a single building in a farm or breeding facility). Note that there was never more than 1 species of *Dermanyssus* per nest.

Note: We used the bird classification according to Peterson (2008), except for the blue tit, which we referred to as *Parus caeruleus* instead of *Cyanistes caeruleus* (L, 1758), in order to match common use in bird banders.

Restricted study areas

In addition, ten special areas with a rather small diameter (<3 km) allowed us testing several different nests used by a single bird community. These restricted study areas are described in Table 1 and indicated by the following acronyms: CB, Ecop, ENVL, HIR, JBO, LB, MOL, RQ in France, IL in The Netherlands, BMOC in the USA.

Nests' analysis

Nests were analysed using a method described by De Lillo (2001) except that no sodium hypochlorite was added to the water solution to wash the stack of sieves and that the sieves had a somewhat different mesh width (top to bottom: 2500, 1400, 180, 100 μm).

Molecular analysis

DNA was extracted from individual mites following a protocol that preserves an intact cuticle for voucher preparation and microscopic observation. Of each sampled population, 2 or 3 mite individuals have been sequenced. A 700–800 bp amplicon of mt-COI gene was isolated by PCR, depending on primer pairs used (i.e. on concerned species, cf. Table 2 and Appendix for EMBL database accession numbers), and then sequenced. PCRs was performed in either a Biometra TGradient or a MWG AG Biotech Primus 96 plus thermal cycler in typical buffer containing 2 μl of template DNA, 2.5 units of Taq polymerase, 10 nmol of dNTPs, 20 pmol of each primer and 3.4 mM $MgCl_2$. After an initial denaturation step (95°C) for 10 min, followed by 40 cycles of 20 s at 95°C (denaturation), 30 s at 52°C (hybridization), and 90 s at 72°C (extension). A final extension step was carried out for 10 min at 72°C. Several primers designed for amplification of DNA from various species are listed in Table 2 and were choosen to perform PCRs under the same conditions.

Negative and positive controls were run with each round of amplification. PCR products were checked by electrophoresis in a 1% agarose gel. Depending on the brightness of the band either additional PCRs were run on the original template or reamplifications of the

Table 1 Description of restricted study areas (cf. "Materials and methods")

Site acronym	Location	Site characteristics	Host	Number of nests tested	Number of nests with *Dermanyssus*	Remarks
ENVL	France, 69	10 nestboxes occupied by great tits or blue tits on the Campus of the National Veterinary School of Lyon from 2007 to 2008	*Parus* sp. (Paridae)	11	6	*D. longipes* in 1 nest of 6 tested in 2007; *D. longipes* in 3 nests of 5 tested in 2008
Ecop	France, 42	Natural protected area Ecopole with several dozens of nestboxes, sampled in automn 2005, summer 2006 and winter 2008	*Parus major* (Paridae), *Phoenicurus ochruros* (Muscicapidae)	15	7	*D. carpathicus* found only from *Parus* sp.—In 2006 and 2008.
RQ	France, 42	House, small garage with small hen house and girder with a redstart nest and two tit nests in a natural protected area at c. 950 m altitude (sampled in 2006, 2007 and 2008)	*Parus major* (Paridae), *Phoenicurus ochruros* (Muscicapidae)	5	5	*D. carpathicus* abundant in redstart nest in 2006, 2007and 2008. Also found in a tit nest sampled in 2006. *D. gallinae* in hen house (Pop. 8004)
MOL	France, 69	A small farm housing sheep, calves, rabbits, dogs and cats, and chickens in four layer hen houses, and with wood girders, housing many swallow nests	*Hirundo rustica* (Hirundinidae)	3	0	Only one specimen of *D. gallinae* (Pop. 8005) collected from hen house in spring-summer 2008 (several liters of litter analyzed)
HIR	France, 85	House with 4 swallow nests in three different rooms	*Hirundo rustica* (Hirundinidae)	6	5	*D. hirundinis* collected in 2 successive years (before nesting in winter 2007, after nesting in spring 2008)
CB	France, 01	Small old farm housing calves, cats and also free-renge chickens according to the production procedure called "AOC Poulets de Bresse", and many swallow nests	*Hirundo rustica* (Hirundinidae)	8	3	Spring-summer 2008. *D. hirundinis* present in swallow nests. Many *D. gallinae* in wooden chicken cages (Pop. 8012)

Table 1 continued

Site acronym	Location	Site characteristics	Host	Number of nests tested	Number of nests with *Dermanyssus*	Remarks
JBO	France, 13	15 plots (maintained with chemical, organic or alternative methods) of apple/pear orchards with one nestbox with great tits every 50 meters in each row of trees	*Parus major* (Paridae)	62	5	Summer 2007. One, three and no nests with *D. gallinae* in resp. alternative control plot (Pop. JBO27), organic plot (Pop. JBO51, JBO46, JBO56) and chemical control plot. One nest with *D. carpathicus* in organic plot (Pop. JBO59).
IL	The Netherlands, Groningen	33 nestboxes occupied by European starlings (populations IL213, IL227, IL302, IL202)	*Sturnus vulgaris* (Sturnidae)	33	32	Summer 2007. *D. gallinae* abundant in many nests.
BMOC	USA, Michigan	4 nests with different birds from the campus of a school in Michigan	*Tachycineta bicolor* (Hirundinidae), *Parus atricapillus* (Paridae), *Troglodytes aedon* (Certhiidae)	4	3	Autumn 2007. *D. hirundinis* rather abundant.
LB	France, 45	Several swallow nests from a goat farm	*Delichon urbica, Hirundo rustica* (Hirundinidae)	10	9	Summer 2006 and 2007. *D. gallinae* present in some nests, *D. hirundinis* in some others.

Table 2 Primer sequences

Primer sense	Primer name	Sequence 5′-3′
Forward	CO1RQF1	GAAAGAGGAACAGGAACAGG
	CO1LCF	GAAAGAGGAGCAGGCACTGG
	COF1bis	CTGCACCTGACATGGCTTTCCCAC
	CO1F4	CACCTGACATGGCTTTCCCACGAT
	RhipiCOIF	CGAATAAATAATATAAGATTTTGA
	SKPOFa2diagF1	CTTTTTAGATCTTTAATTGAAA
Reverse	COIGOR	GTTGGAATtGCAATAAT
	RQ-COI-R	CCAGTAATACCTCCAATTGTAAAT
	ObCOIF-rev	GTGGGAATHGCAATAAT
	TyphloCOIR	GCTAATCAAGAAAAAATTTTAAT

Primer pairs used in present study for the amplification of mt-COI according to species indicated into brackets: CO1RQF1 + RQ-COI-R (*D. carpathicus*, *D. hirundinis*, *D. longipes*), SKPOFa2diagF1 + RQCOIR, CO1F4 + RQCOIR, COF1bis + RQ-COIR, COF1bis + ObCOIF-rev, CO1LCF + RQ-COI-R (*D. gallinae*), RhipiCOIF + TyphloCOIR (outgroups), CO1LCF + COIGOR (*D. apodis*)

original PCR product were performed. PCR products were sequenced by Genoscreen (France, Lille) using a 96-capillary sequencer ABI3730XL.

Phylogenetic reconstruction

Sequence data was aligned using MAFFT (Katoh et al. 2005) with the L-INS-i iterative refinement option on the MAFFT server at http://align.bmr.kyushu-u.ac.jp/mafft/online/server/. MAFFT with the L-INS-i option has shown to be the most accurate and consistent method for the alignment of sequences (Wilm et al. 2006; Carroll et al. 2007).

The alignment of 558 bp from cytochrome oxidase I (mt-COI) was analysed under several optimality criteria: (1) Parsimony using PAUP* 4.0b10 (Swofford 2001) to build tree(s) with TBR branch swapping and 10,000 random additions saving all most parsimonious trees, (2) Bayesian inference using the computer program MrBayes v3.1.2 (Huelsenbeck and Ronquist 2001; Ronquist and Huelsenbeck 2003) with the GTR + Γ+i model of evolution chosen in the computer program MrModeltest (Nylander 2004) that was run for 5,000,000 generations. Burnin was determined based on stationarity being reached. The clade notation taxon^{+} indicates the clade including the taxon at its base and all subsequent taxa included in the same clade (De Souza Amorim 1982).

Bioassays

In order to get some data complementary to phylogenetic information, some bio-assays were performed. Eight different bioassays have been performed in order to compare the behaviour of four *Dermanyssus* species. Note that most of these bioassays are rather tentative (method not completely validated), provide limited information, and are often not mutually comparable. However, some of the data obtained add interesting elements to the discussion.

The first two (comp1 and 2) involved a permanent or intermittent contact of mites with one canary, during a period of several weeks (cf. Table 3). A single canary was placed in a plastic box, provided with several apertures on the bottom and sides covered with a filter tissue for aeration, and with an eating and a drinking trough that can be filled from the exterior of the box. This enclosure was placed into a large bowl filled with water and a drop

Table 3 Some data on development of 3 *Dermanyssus* species on canaries, obtained from long-time bioassays (referred to in text as comp1 and comp2)

Bio-assay		Host	Mite species	Mite population	Total number of mites tested	Duration of mite-bird contact	Total number of living mites collected
comp1 (permanent contact)	a	Canary	*D. gallinae*	SK	20	98 days	205
	b	Canary	*D. longipes*	PAS	40	98 days	0
	c	Canary	*D. longipes*	PAS	>200 (nest residu)	>100 days	12
comp2 (intermittent contact: during 12 h every 2 weeks, all along 1 year)		Canary	*D. carpathicus* + *D. gallinae*	RQ + SK + Fal	>200 (nest residu + mites)	ca. 24 days	>1000 D.gal.; 8 D.car.

Bioassay comp1 is split into three lines (a, b and c), which correspond to three boxes, each containing one canary, and which have been handled simultaneously and in the same conditions. Information about populations is to be found in the appendix Mite population's codes correspond to population codes in Appendix of tension-active agent. Water and food were added regularly and the canary was allowed to move, feed and drink ad libitum. Mites introduced into the enclosure cannot escape and no mites from outside can get in. Moreover, the escape of mites from the enclosure into the surrounding water was regularly checked, which represents less than 10% of final count. At the end of the period, the litter and plastic box were rinsed and treated following the method of De Lillo (2001). Mites were observed and counted using a stereoscopic magnifying glass.

The 5 other bioassays (comp3 to 8) consisted of short periods of contact between the mites and bird (several hours at a time) in glass containers (cf. Table 4). One small bird (canary, hen chick, duck chick) was placed within the container, which was placed into a bowl filled with water and a drop of tension-active agent. Next, mites were added (an aggregate consisting of an unknown number of individuals, or–in some cases–of a known number of individuals) and the whole device was kept in darkness (incubator at 25°C). In case of known number of handled mite individuals, mites were collected at lab using a vacuum pump and 10 μL filter tips (tips' filters are permeable to air, but retain mites). Tips were closed with some parafilm for storage and broken just before being introduced in the glass container with the canary at the beginning of the bioassay. After several hours, birds were removed, devices were examined and engorged mites were isolated in ELISA microplates, covered with extended parafilm (one small cut above each well, insertion of a single mite using a fine wet brush, obturation of cuts using a small piece of extended parafilm). Cues that provide information on developmental progress (exuvia, eggs) were recorded using a stereoscopic magnifying glass at d + 4 and noted down as follows: 1 exuvia (protonymph, deutonymph) or 1 laying (1 to several eggs) (adult female) in a well indicates that the isolated individual has developed.

Results

Inventory of *Dermanyssus* species from wild and domestic bird

Overall, 27 bird species distributed across nine different bird orders were examined. Among wild avifauna, 334 nests of wild birds, representing 25 bird species, distributed across seven bird orders and 31 families, have been analysed (Table 5). Note that the goal here is not to get an overview of prevalence and that this study is not an epidemiological one. Here, we report the results of our explorations on host specificity, based on simple records on a batch of samples obtained from various bird species and places. That is the reason why even some bird groups with only few nests examined are noted. Anyway, the focus will be on bird groups with a significant number of samples analysed (Fig. 1).

Moreover, hundreds of wild birds have been examined including Picidae, Alaudidae, Coraciidae, Paridae (adding one more bird species to our study: *Coracias garrulus* L., 1758).

Additionally, several dozens of mites have been collected from layer hen farms (hence one more species included in our study: *Gallus gallus* L., 1758) and some breeding facilities (chicken, pigeons, canaries, included in Fringillidae in Table 5).

Dermanyssus was present in nests of 10 wild bird families (Table 5). Moreover, one additional bird family parasitized by *Dermanyssus* was found by direct on-host sampling (during bird banding and bird care activities): *C. garrulus* (Coraciidae: Coraciiform) (Populations ROL1 and ROL2 in Appendix).

Overall, in France and in the Netherlands, 5 species of *Dermanyssus* have been isolated from nests and birds: *D. gallinae* including a special lineage, which may represent a cryptic

Table 4 Engorgment and development of 3 *Dermanyssus* species compared using three host species and short-time bioassays (referred to in text as comp3 to comp8)

Assay		Host	Mite species	Mite population	Total number of mites tested	Duration of contact mite-bird (hours:minutes)	Number of engorged mites collected	% of live mites that were engorged upon collection	Total number of mites recovered (including those not engorged and submerged into water)	Number of development cues (d + 4)	% of mites having developed at d + 4 (mites engorged and collected alive)	Number of individuals dead after d + 0	Remarks
comp3		Duck chick	*D. gallinae*	SK	unknown	12:00	120	–	NE	120	100	NE	
comp4	a	Canary	*D. carpathicus*	RQ	unknown	04:40	3	–	NE	0	0	3	All mites were dead when collected
	b	Canary	*D. gallinae*	8,010	unknown	04:40	51	–	NE	27	53	1	
	c	Canary	*D. hirundinis*	HIR5	unknown	04:40	45	–	NE	15	33	1	
comp5	a	Hen chick	*D. carpathicus*	RQ	154	05:30	0	0	0	0	–	0	
	b	Hen chick	*D. gallinae*	8,010	154	05:30	33	54	97	49	59	7	
	c	Canary	*D. carpathicus*	RQ	154	06:30	0	0	0	0	–	0	
comp6	a	Canary	*D. gallinae*	8,010	154	11:45	29	19	86	24	83	2	
	b	Hen chick	*D. gallinae*	8,010	154	11:45	50	39	90	46	77	4	
	c	Hen chick	*D. gallinae*	8,010	154	11:45	30	19	61	10	33	14	
comp7	a	Canary	*D. gallinae*	8,010	154	05:00	31	20	78	22	71	0	
	b	Canary	*D. gallinae*	8,010	154	05:00	22	14	70	10	45	3	
	c	Hen chick	*D. gallinae*	8,010	154	05:00	10	6	80	4	40	3	At least 77 individuals (N1) remaining within the filter tip, likely due to diarrhea produced by chick

Table 4 continued

Assay	Host	Mite species	Mite population	Total number of mites tested	Duration of contact mite-bird (hours:minutes)	Number of engorged mites collected	% of live mites that were engorged upon collection	Total number of mites recovered (including those not engorged and submerged into water)	Number of development cues (d + 4)	% of mites having developed at d + 4 (mites engorged and collected alive)	Number of individuals dead after d + 0	Remarks
	d Hen chick	*D. gallinae*	8,010	154	05:00	48	31	147	27	56	2	At least 77 individuals (N1) remaining within the filter tip, likely due to diarrhea produced by chick
comp8	a Canary	*D. carpathicus*	RQ	>150	12:00	8	<5%	33	–	–	–	The 8 engorged individuals found dead in liquid canary's droppings.
	b Canary	*D. gallinae*	8,010	>150	12:00	82	<55%	95	–	–	–	

Subdivisions in comp4 to comp8 (a, b, c, d) correspond to the different glass containers involved in the bioassay, containing each one small bird, and which have been handled simultaneously and in the same conditions. Recorded development cues are exuviae and eggs. Information about populations is to be found in the appendix

NE not estimated

Table 5 Number of nests analysed per bird family and occurrence of genus *Dermanyssus* based and on present field data and on literature

Bird order	Bird family	Nests analysed in present study	Mite populations from farms or breeding facilities	Wild bird individuals on which mite populations have been directly found	*D. alaudae*	*D. apodis*	*D. americanus*	*D. antillarum*	*D. brevirivulus*	*D. brevis*	*D. carpathicus*	*D. chelidonis*	*D. faralloni*	*D. gallinae*	*D. gallinoides*	*D. grochovskae*	*D. hirsutus*	*D. hirundinis*	*D. longipes*	*D. nipponensis*	*D. passerinus*	*D. prognephilus*	*D. quintus*	*D. rwandae*	*D. transvaalensis*	*D. triscutatus*	*D. trochilinis*	*D. wutaiensis*
Apodiform	Apodidae	52		7		F		L										L						L				
	Trochilidae																										L	
Ciconiiform	Accipitridae	3						L						F				L										
	Falconidae	1																										
	Alcidae												L															
	Ardeidae	2																										
	Ciconiidae	1																										
	Hydrobatidae												L															
Columbiform	Columbidae	9	2	1										L				L										
Coraciiform	Coraciidae			2										F														
	Meropidae													L				L										
Galliform	Phasianidae		18											M														
Gruiform	Gruidae	1																										
Passeriform	Alaudidae	14			L				L	L																		
	Certhiidae	1																M										
	Cinclidae	1																										
	Corvidae	1														L												
	Fringillidae	4	3				L					L		M							L	L						
	Hirundinidae	39										L		L				M				L			L	L		
	Laniidae																	L										
	Motacillidae																	L										
	Muscicapidae	23									M			L				L	L		L	L						
	Paridae	124									M	L		M				M	M		L							
	Passeridae	30					L	L						L				L	M		L							L
	Sittidae						L							L				L										
	Sylviidae	1												L				L										
	Vireonidae																	L										
Piciform	Picidae	1		1										M	L	L	L	L		L	L	L	L					
Strigiform	Strigidae	2					L							L				L										
	Tytonidae	1												F														
Sturniform	Sturnidae	33						L						M				L	L		L							

Source of literature references: Roy and Chauve (2007) for a review including 23 *Dermanyssus* species, Knee (2008) for *D. diphyes*, Roy et al. (2008) for *D. apodis* and some data concerning *D. longipes*. Taxonomy of birds is according to Peterson'website

L data from literature only, *F* data from our field work only, *M* mixed data from literature and from our field work

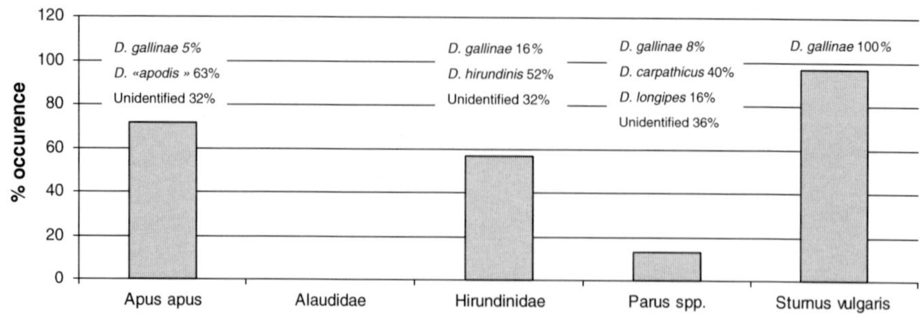

Fig. 1 Percentage of occurrence of *Dermanyssus* in nests of the five bird groups under test. Above each column the percentage of identified species among *Dermanyssus* individuals is given

species (here referred to as L1), *D. hirundinis, D. carpathicus* (Zeman 1979), *D. longipes* (Berlese and Trouessart 1889) and *D. apodis* Roy et al. 2008).

The number of nests analysed per bird group varied considerably, but a substantial number of nests were available in five bird groups and these represented together *ca.* 80% of all nests analysed. These groups are scrutinized below.

Apus apus, *the black swift*

Of the 52 nests analysed, 79% contained *Dermanyssus* individuals. Of these *Dermanyssus*, 32% were not identifiable at the species level because of their poor preservation condition. Of the remainder, 63% was identified as *D. apodis* and 5% as *D. gallinae*. In addition to *Dermanyssus* spp. collected from nest material, 18 individuals belonging to *D. apodis* were collected directly from the host (on several bird individuals), seven of which were on chicks in the nest and 11 on adults or on a fledged young (found far from nest).

Sturnus vulgaris, *the European starling*

Only one of the 33 nests analysed did not harbour any *Dermanyssus* individuals. Of the 32 others, four populations have been sequenced (mt-COI), which all appeared to belong to *D. gallinae*. Moreover, haplotypes found in these populations appeared to be very close to each other.

Parus major *and* P. caeruleus, *the great tit and the blue tit*

On the whole, 120 nests have been tested in the two species of tits, 62 of which originated from apple and pear orchards.

Of all these nests, 21% contained some *Dermanyssus* individuals, but in orchards, 8% provided some *Dermanyssus* individuals, versus 34.5% *in natura*. Moreover, almost all individuals collected from orchards were identified as *D. gallinae*, whereas mites from nests sampled *in natura* were identified as *D. carpathicus* or *D. longipes*. Moreover, several hundreds of bird individuals have been examined, without finding any individual of *Dermanyssus* spp.

Alaudidae, the larks

Nests of three species of Alaudidae have been examined, one of these species studied most extensively was *Melanocorypha calandra* L., 1766. No *Dermanyssus* were found in any of the nests.

Moreover, no more *Dermanyssus* have been found on any of the hundreds of bird individuals (mainly *Alauda arvensis* L. 1758) examined from two different regions in France around 800 km apart (Drôme, department 26, Pas de Calais, department 62). In particular, 50 individuals caught by a hunter in Drôme were closely examined, then submerged in water and analysed using the De Lillo's method, but there was no recovery of any *Dermanyssus* individual.

Hirundinidae, the swallows

Nests were obtained from two species of hirundinids: *Delichon urbicum* L., 1758 and *Hirundo rustica* L., 1758. Of the 42 nests analysed, 58% provided *Dermanyssus* individuals (mostly *D. hirundinis*). Moreover, the guild of bird parasites appeared to be more diverse, with several groups in addition to *Dermanyssus*, such as fleas (Insecta: Siphonaptera), chewing lice (Insecta: Mallophaga) and an individual belonging to *Myonyssus* sp. (Mesostigmata: Laelapidae).

Phylogenetic analysis

The data matrix consisted of 558 characters from a coding region of cytochrome oxidase subunit 1 (COI), of which 216 were parsimony-informative. The 78 included taxa, corresponding to mite populations, include 3 distant outgoups, 2 species of the *hirsutus*-group, and 73 populations of *gallinae*-group (groups classified according to Moss (1978)). Note that different haplotypes within a single population was detected in only two populations (RQ-Mes and 8006, 2 haplotypes of a single species detected in each), which come, respectively, from one nest and one farm. Parsimony analysis recovers 1000 most parsimonious trees ($L = 775$, $CI = 0.5316$, $RI = 0.8879$, Fig. 2), with 55 distinct haplotypes of *Dermanyssus* (53 of *gallinae*-group). Bayesian analysis resulted in a topology similar to MP, slightly more resolved concerning internal relations of the *D. apodis*$^{+}$ clade (Fig. 3).

The same groupings appear in analyses using a subset of taxa and a combined matrix concatenating a region of rRNA 16S, a region from rRNA 18S to 28S, including ITS1, 5.8S and ITS2, as well as the COI region (Roy et al. 2008).

Populations sharing haplotypes

Populations, as defined above, come from different nests or different farms or breeding facilities, from different places. Most of mites obtained come from France (cf. Fig. 4 in Appendix), some additional samples come from Belgium, Denmark, The Netherlands, Spain, Italy, Norway, Poland and the USA (cf. Appendix). Each population is at least 3 km apart from each another, except in the cases of restricted study areas (Table 1) in which some nests stood several hundreds meters apart from each other.

Some haplotypes are to be noticed, as these were found in several populations of *D. hirundinis* and *D. gallinae* from single restricted study areas. For example, a single haplotype occurred in populations from areas BMOC, CB and IL, (except in two populations, which had haplotypes differing in only one (IL 227) or two (CB3) nucleotide

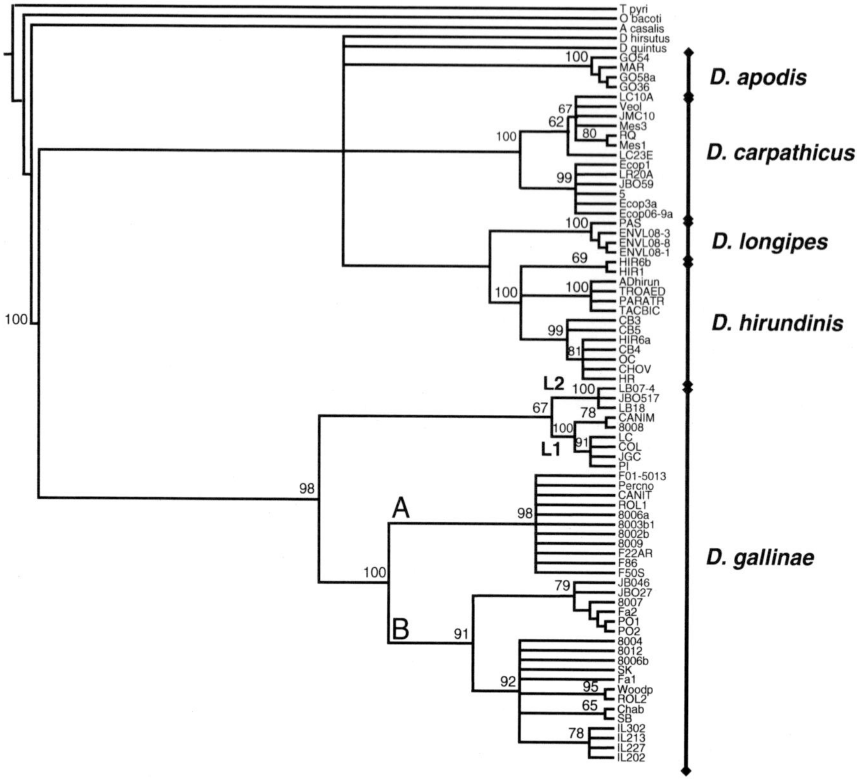

Fig. 2 Maximum Parsimony analysis. PAUP 4.0. Strict consensus of 1000 most parsimonious trees. Description of these 1,000 trees: $L = 775$, $CI = 0.5316$, $RI = 0.8879$. *Numbers* at nodes refer to bootstrap percentages for 1,000 replicates. Two lineages discussed in text are labeled L1 and L2. Two clades discussed in text are labeled A and B

substitutions). On the contrary, populations of *D. carpathicus* and *D. longipes* from single restricted study areas provided similar but slightly different haplotypes, as observed for populations from areas RQ, Ecop, ENVL. This suggests that populations' intermingling occurs, but the extent differs from species to species. The existence of single haplotypes in some restricted study areas suggests a single source. Possibly, starlings from restricted study area IL and swallows from CB have been infested by only one population of *D. gallinae* and *D. hirundinis*, respectively, whereas several infestation events may have occurred in areas RQ, Ecop and ENVL, with *D. carpathicus* and *D. longipes*.

Within the *D. gallinae* clade, 9 different populations from various geographic origins in France share a single haplotype (layer hen farms: 8009, 8002b, 8003b1, 8006a, F01-5013, F50S, F86, F22AR; wild bird: Percno; cf. Fig. 3). This group is labelled A in Figs. 2 and 3.

Observed host range

Comparison between literature and field data

Table 5 provides an overview of the distribution of *Dermanyssus* species across bird hosts using published literature data (based on morphological diagnosis of mites) and our field

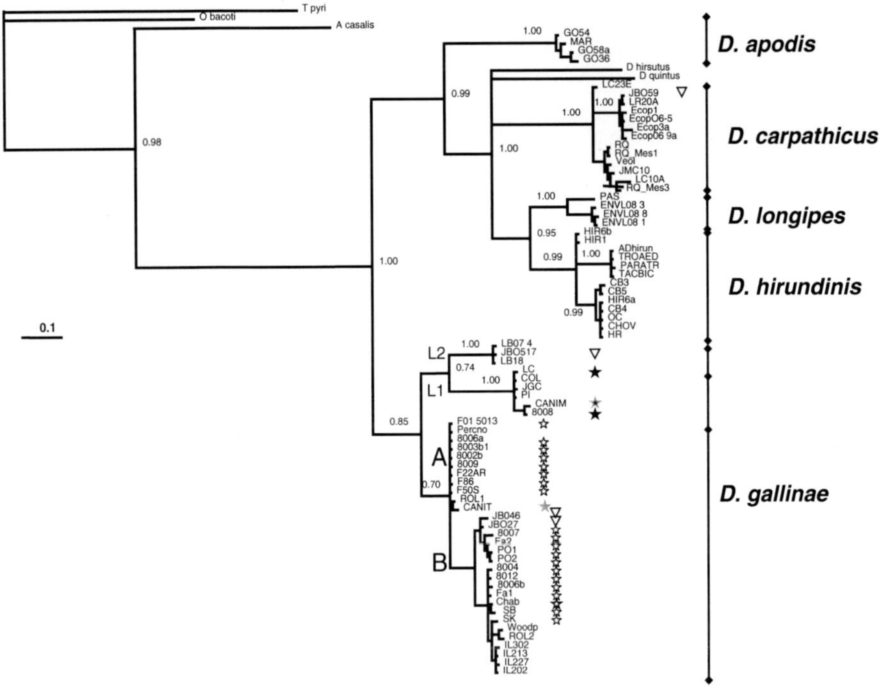

Fig. 3 Bayesian analysis. MrBayes v3.1.2, GTR + Γ+i model of evolution for 5 × 10^6 generations. Numbers at nodes refer to Bayesian Posterior Probabilities. Additional symbols indicate the type of anthropogenic ecosystem: "★" pigeons breeding facilities, "★" canary breeding facilities, "☆" layer hen or chicken houses, "▽" apple/pear orchards. Populations without any of these symbols have been collected in natura. Two lineages discussed in text are labeled L1 and L2. One clade discussed in text is labeled B. Group A corresponds to clade A in Fig. 2, and groups together populations sharing the same haplotype

data (based and on morphological and molecular diagnosis of mites). It includes data on 111 bird species (two of which undetermined: *Parus* sp. and *Passer* sp.), distributed over 9 bird orders and 31 bird families. Of these 111 species, data on 69 bird species were derived from the literature only, data on 17 bird species stemmed from literature and our field data and data on 25 bird species originated only from our field samples, which included rather variable numbers of nests per bird species.

Twenty-five mite species are currently included in the genus *Dermanyssus.* Of these 25 species, five have been found in our field samples (*D. carpathicus, D. gallinae, D. hirundinis, D. longipes, D. apodis*).

D. carpathicus was found in association with four passeriform bird species of two different genera [*P. major* and *P. caeruleus, Phoenicurus phoenicurus* (L., 1758) and *Ph. ochruros* (Gmelin, 1774)]. These host genera were known from the literature.

D. gallinae was found in various, distant bird groups. It was previously recorded from 25 bird species, 4 of which were in our field samples [*G. gallus, Serinus canaria* (L., 1758), *P. major, S. vulgaris*]. Our field data also provided records of three bird species [*Dendrocopos major* (L., 1758), *Neophron percnopterus* (L., 1758) and *C. garrulus*] that are new as hosts of *D. gallinae.*

D. hirundinis was found exclusively and frequently (at least 30%) in nests of two species of Hirundinidae, the type host family in France. Since it was previously recorded from almost 40 different bird species distributed in nine different bird orders, we would have expected a wider host range. Recently, some individuals from the restricted study area BMOC (Table 1) were found not only in nests of swallows (Hirundinidae), but also in nests of tits (Paridae) and wrens (Certhiidae) in the USA (cf. "Host switches via nest sharing of host birds").

D. longipes has been found in a nest of *Passer montanus* and in several nests of *Parus* sp. Its type host was *Passer domesticus* (L., 1758). Moreover, some mites found by Brännström et al. (2008) from several species of Muscicapidae (Passeriforms) provided an ITS sequence similar to that obtained from our populations of *D. longipes* (Roy et al. 2008).

Host transfer of populations from four Dermanyssus *species*

To assess the ability of four *Dermanyssus* species to feed on hosts other than ones they are associated with, bioassays were carried out. These bioassays were not repeated or standardized enough to be dealt with statistically. At best they may give a hint as to the ability to feed and develop on the new host (cf. Tables 2, 3).

D. gallinae A strain of *D. gallinae* named SK and cultured on hens for the last ten years was, transferred to ducks and canaries, where they readily fed and successfully reproduced. Another population of *D. gallinae* (8010) also showed such an ability to feed on canaries under laboratory conditions, immediately after having been collected from a layer hen farm (after a starvation period of 4 days). This population shares the haplotype of group A (cf. above), *ie* provides a haplotype very common in French layer farms.

D. longipes and D. carpathicus *D. longipes* and *D. carpathicus* did not reveal an ability to feed on a new host to the extent observed for *D. gallinae*. *D. longipes* was unable to develop on a canary in the laboratory (long time bioassay). A population of *D. carpathicus* caught from a nest of *P. ochruros* (Gmelin 1774) appeared to maintain itself during about one year on a canary, since brilliant red individuals were regularly noticed. However, short time bioassays with canaries as the new host suggested a very different behaviour in *D. carpathicus* than in *D. gallinae*. In the former species, most individuals released were not recovered and there were only a few engorged mites observed, whereas in the latter species, simultaneously under similar conditions, most individuals were found aggregated and engorged.

D. hirundinis Although tested only once in bioassays, *D. hirundinis* did not seem to be different from *D. gallinae* in its ability to feed and develop on canaries as a host.

Discussion

Two main clades appear in the species of *Dermanyssus* tested here. One clade includes *D. carpathicus, D. hirundinis, D. longipes,* and *D. apodis*. The second clade includes the various lineages of *D. gallinae*. Within the *D. gallinae* clade, two lineages are strongly separated from the others.

D. gallinae: several lineages, some more specialized than others

The two lineages that stand out as strongly isolated from each other and from the other lineages, form a sister group to all other *D. gallinae*. All analyses provided strong support for the monophyly of populations of L1 (Fig. 3) that were collected mainly from pigeons (domestic pigeons from breeding facilities and a nest sampled in a town) and stem from distant geographic areas. The monophyly of lineage L2 also receives strong support from the analyses shown in Figs. 2 and 4, but is not so strongly supported in other single gene analyses (Roy et al. 2008). Populations in this lineage were only found three times in wild Passeriformes.

The sister group of the clade L1 + L2 groups together populations found in industrial layer hen farms and some other populations (67% bootstrap in MP, 0.74 Bayesian Posterior Probabilities in Bayesian analysis). Within this clade, two major groupings become manifest, that are also found in wild avifauna: one grouping concerns *D. gallinae* populations in hen farms, in nests of *N. percnopterus* and on *C. garrulus* (group A, Figs. 2, 4) and the other grouping concerns *D. gallinae* populations in hen farms and in nests of *S. vulgaris, P. major* and *D. major* and on *C. garrulus* (clade B, cf. Figs. 2, 4). All populations in group A except ROL1 and CANIT provided the same haplotype, even though these populations are of quite different geographic origin (cf. Appendix). In clade B, approximately one haplotype per population is present, but in some cases there were haplotypes with very small differences: for instance the haplotype of PO1 and PO2 (layer hens from Poland) is very close to that of Fa1 (layer hens from Norway), differing by a single nucleotide substitution.

Assessments on host specificity using observed host range and ecological observations

The clade *D. apodis*$^+$ had a much narrower host range than *D. gallinae:* only Passeriformes represent hosts for *D. carpathicus, D. longipes* and *D. hirundinis,* Piciformes for the *hirsutus*-group and Apodiformes in the basal *D. apodis,* whereas *D. gallinae* was isolated in eight different bird orders.

*Host spectrum enlarging within the most synanthropic clade (*D. gallinae*)*

L1 and L2 appear to be more specific than their sister clade. L1 was mainly found in association with pigeons and L2 with bird species of two different passeriform families. For L1, there were only two records from other bird groups, but these could well be cases of mites that do not actually infest these birds: one dead and dried individual was found in a nest of a predatory bird [*Tyto alba* (Scopoli 1769)] and another single individual was found dead in a nest of black swifts sampled in the town of Nîmes. Since in the town environment pigeons and swifts tend to compete for nesting places, the latter individual may have stemmed from a pigeon host. For these reasons, L1 could be specific to pigeons. L2 was never found in farms or other anthropogenic environments, whereas L1 was found in facilities for pigeon breeding.

The remainder of the *D. gallinae* populations showed very little divergence in mt-Co1 (only 19 haplotypes among 30 populations, pairwise divergence percentage 1–6%, vs. 8–12% between these populations and L1 or L2), were mainly collected from hens (Galliformes), but also from completely different bird groups (Coraciiformes, Piciformes, Passeriformes, Ciconiiformes). Moreover, some lab bioassays succeeded in feeding mites of different *D. gallinae* populations, freshly caught from hen farms, on canaries and on duck chicks and these mites showed a rate of development similar to that on their original host (Tables 2, 3). These *D. gallinae* populations seem to quickly adapt to new hosts and the large clade to which these populations belong was found exclusively in layer hen farms.

Host switches via nest sharing of host birds

The cases where haplotypes were found to be restricted to certain areas (sometimes with several successive samples, year after year; Table 1) require scrutiny because they may help to get insight into a possible mode of dissemination of *Dermanyssus* mites. For example, 2 sites with *D. carpathicus*, one with *D. longipes*, two with *D. hirundinis* and one with *D. gallinae* led us to suggest (1) that a single species was present in association with 1 (or 2) bird species in most areas that had a diameter compatible with the home range of birds under consideration, and (2) that mixing of mite populations occurs in restricted study areas in all *Dermanyssus* species under test. Thus, often a single haplotype was found per restricted study area, or haplotypes differing by only a few nucleotide substitutions. In 2 nests, however, two different populations were noted in the same *Dermanyssus* species (HIR6a and HIR6b both belonging to *D. hirundinis,* RQ-mes1 and RQ-mes3 both belonging to *D. carpathicus*). The concerned bird species (tits in genus *Parus*, starlings, swallows) are known to spend all time in a restricted area throughout the year and to rebuild their nest upon an older one, either their own or that of another (Caparros, Bouvier, Personal communication). Thus, there is ample opportunity for the *Dermanyssus* mites to switch from one host to another in case their hosts share nesting places, as suggested by Valera et al. (2003).

Contact between nest and bird seems to be absolutely necessary. This mode of dissemination via nest sharing is supported by our results in that different host ranges were noted in *D. hirundinis* between France (populations HIR6a, HIR6b, HIR1, CB3, CB4, CB5, OC, CHOV, HR) and the USA (populations ADhirun, TROAED, PARATR, TACBIC) (cf. Figs. 2, 3, and appendix for host affiliation). Indeed, this species has been isolated exclusively from *H. rustica* and *D. urbicum* (Hirundinidae) in France, whereas it was found from three different passeriform families in the USA (Hirundinidae, Certhiidae, Paridae). But the American hirundinid species tested is not present in France (*Tachycineta bicolor* (Vieillot 1808)) and its ecology strongly differs from the 2 tested French species: *T. bicolor* or the Tree Swallow is a cavity nesting bird and often uses nestboxes in the USA, as do the two other American host species (*Poecile atricapillus* L., 1766, the Black-Capped Chickadee and *Troglodytes aedon* Vieillot, 1809, the House Wren). These three bird species are often found sharing the same nestboxes in the USA (O. Dehorter, Personal communication), whereas none of the two tested French hirundinid species are found sharing nestboxes in France.

This opportunistic behaviour of *D. hirundinis* is also observed in our bioassays. Individuals of *D. hirundinis* directly sampled from a fresh nest of *H. rustica* did not show differences with individuals of *D. gallinae* in engorgement and development on canaries as hosts. The apparent host specificity observed in France is therefore likely the result of ecological and/or geographic factors. Alternatively, there may be genetic differences between the French and American populations tested, as indicated by small differences in COI sequences.

Evidence strongly suggests that dissemination of mites happens in cases of nest sharing between congeners and different species. The ability to switch from one host to another at a distance of several meters seems to be greatly reduced at least in some of these species. As indicated by scrutiny of the population in the restricted area CB (Table 1), some *D. gallinae* may rather starve for several weeks to a few months, than to venture bridging the distance to new hosts only metres away. For example, in absence of chicken, *D. gallinae* individuals present in large numbers in chicken cages do not appear to move to nests of swallows nearby. In these swallow nests, only *D. hirundinis* was found!

The nest seems to be a reservoir and the bird host could be the carrier. Note that all *Dermanyssus* individuals directly collected on flying birds in the present study (cf. Appendix) were adult females. Is this a stage adapted to dissemination by phoresy? Note also that, Flechtmann and Baggio (1993) reported one case of *D. gallinae* phoretic on a beetle, and the isolated individuals were adult females.

Synanthropicity and nest microenvironment

The micro-environmental conditions of farms and breeding facilities are likely to differ from those in wild bird nests. For example, these anthropogenic environments harbour large numbers of bird individuals in a small area, provide relatively regular temperatures and humidity, are usually exposed to some pesticides and provide hosts during most of the year (layer hen houses are usually bird-free for less than 2 months per year). On the contrary, in wild bird environments, even in cases of bird colonies, the number of bird individuals and the area they occupy is much smaller than in farms, and temperature and humidity are much more variable during the year and even during the day, than in farms. Also, pesticide products are typically absent (except in the case of orchards in restricted study area JBO) and the host is available only during a limited period, i.e. the breeding period of their host in spring-summer, and sometimes winter nights (implying absence of bird host for several months in autumn and summer).

The two main clades in the *Dermanyssus* species tested exhibit conspicuous adaptations to their microenvironments. Clade *D. apodis*$^+$ was never found in any farm or breeding facility, but clade *D. gallinae* was present in poultry farms (layer hens, chickens) or breeding facilities for canaries, other Fringillidae and pigeons, where they usually proliferate. Moreover, the latter clade was found in agroecosystems such as orchards. Taken together, it can be concluded that the clade *D. gallinae* is unique in harbouring synanthropic populations (Fig. 3).

Role of the ecosystem in the nest environment

The nest provides a specific environment shaped by various organisms together forming a micro-ecosystem. First and foremost are the birds occupying the nest. They bring nest-building material and release waste products, but they may also remove them. So nest building and nest hygiene may be determinants of the nest as a biotope for *Dermanyssus* spp. In wild birds, *D. gallinae* was found proliferating only in pigeon nests (L1) and starling nests (restricted study area IL), which are birds that allow droppings from chicks to dry within the nest. Moreover, *D. gallinae* is commonly found in layer hen farms where hen droppings accumulate around the flock and numerous *D. gallinae* individuals are often found aggregating under dried droppings. In contrast, *D. hirsutus, D. quintus, D. carpathicus, D. longipes* and *D. hirundinis,* in clade *D. apodis* $^+$, were never found in such nests and seem to proliferate only in passeriform or piciform nests which are regularly cleaned by the parental birds. One may wonder how the presence of bird droppings affects species of clade *D. apodis* $^+$ relative to *D. gallinae*.

Another factor affecting the proliferation of *Dermanyssus* spp. could be the presence or absence of pesticide products. Results from nests of restricted study area JBO suggest that *D. gallinae* is better adapted to an orchard agroecosystem than *D. carpathicus:* the latter was found *in natura* (either in natural nest or in nestbox) in 15% of the *Parus* nests sampled in France, whereas only once in fruit orchards (62 nests; i.e. 1,6%). However, *D. gallinae* was found in tit nests only when the nests were located in orchards (Fig. 1).

Finally, various other organisms inhabit nests, these are mostly insects and mites and represent different feeding guilds: bird parasites, predators, microbivores and scavengers. It is interesting to observe that tree-nesting birds generally suffer from attack by *Dermanyssus* spp. and that the breeding success of individual birds strongly depends on the presence of predatory mites that attack *Dermanyssus* spp. (Lesna et al. 2009). The only ground nesting group extensively analysed here are the larks (Alaudidae) and these bird species appear to stand out as the only ones that had no *Dermanyssus* individuals in their nests (Fig. 1). As a rule, these nests contained many species of predatory arthropods (data not shown) and the presence of various predators was inversely proportional to the presence of *D. gallinae*. The case of restricted study area MOL is a good illustration of it, with only one *D. gallinae* individual found among several cubic meters of litter sampled, but several species of predators present in considerable numbers.

Perspectives

To understand the distribution of *Dermanyssus* spp. and lineages over bird species as hosts, co-phylogenetic analysis is of fundamental importance (especially with respect to *D. carpathicus, D. hirundinis, D. longipes, D. apodis* and their bird hosts). However, the presence of a *Dermanyssus* species in the nest of a bird species depends on several ecological factors that may dramatically alter the potential host range. These include bird-related factors, such as nest material selection, nest hygiene, competition for nest sites and nest sharing, but also the presence of various guilds of nidicolous arthropods, including predators of *Dermanyssus* spp. and flying insects that may act as vectors for dissemination of *Dermanyssus* spp., being wingless and therefore less mobile. Given that anthropogenic environments, such as poultry farms, offer an environment that is widely different from that of wild birds, it is to be expected that *Dermanyssus* spp. (especially *D. gallinae*) undergo strong selection to adapt to this environment (*e.g.* by developing pesticide resistance). It is the extent to which these adaptations are decisive for survival in the anthropogenic environment and the extent to which they affect survival in the natural environment, which will determine how much exchange there will be between *Dermanyssus* from wild and domestic birds. Elucidating these adaptations will therefore be an important task for future research.

Acknowledgments Authors would like to warmly thank for help, sampling, advices: O. Bain (Muséum National d'Histoire naturelle (MNHN), Paris, France), Y. Beauvallet (Centre de Recherche sur la Biologie et les Populations d'Oiseaux (CRBPO), MNHN, Paris, France), S. Bonnet (LEGTA, Saint-Genis Laval, France), J. C. Bouvier (INRA unite PSH, Avignon, France), L. Brucy (CRBPO, MNHN, Paris, France), O. Caparros (CRBPO, MNHN, Paris, France), L. Carrier (CRBPO, MNHN, Paris, France), F. Humbert (CRBPO, MNHN, Paris, France), T. Cencek, (Państwowy Instytut Weterynaryjny, Puławy, Poland) M. De Rojas (Universidad de Sevilla, Sevilla, Spain), P. Chabrol (veterinary office, Bourg-en-Bresse, France), J. M. Chavatte (MNHN, Paris, France), O. Dehorter (CRBPO, MNHN, Paris, France), J. Delaporte (Bayer Animal HealthCare, Saint-Ave, France), A. G. Gjevre (National Veterinary Institute of Norway, Oslo, Norway), G. Gory (Museum d'Histoire naturelle, Nîmes, France), N. Guichard (LEGTA, Saint-Genis Laval, France), P. A. Heuch (National Veterinary Institute of Norway, Oslo, Norway), G. Inizan (Bayer Animal HealthCare, Saint-Ave, France), O. Kilpinen (Danish Institute of Agricultural Sciences, Lyngby, Denmark), S. Kreiter (Supagro, Montpellier, France), G. Lallemand (Lycée des Mandailles, Châteauneuf de Galaure, France), B. M. OConnor (Museum of Zoology, University of Michigan, Ann Arbor, USA), B. Paoletti

(University of Teramo, Teramo, Italy), M. Rigaux (IUT A, Université Lyon1, Villeurbanne, France), P. Tavernier (Centre de Soins aux Oiseaux sauvages du Lyonnais, Francheville, France), F. Veau (CORA Ardèche, France), Vincent-Martin N. (CRBPO, MNHN, Paris, France), J. Witters (Institute for Agricultural and Fisheries Research, Merelbeke, Belgium).

Glossary

Mono-/polyxeny Condition of host specificity for a parasite species that needs a single host species/several host species for completion of its development.

Synanthropic Ecologically associated with humans.

Appendix

See Fig. 4 and Table 6.

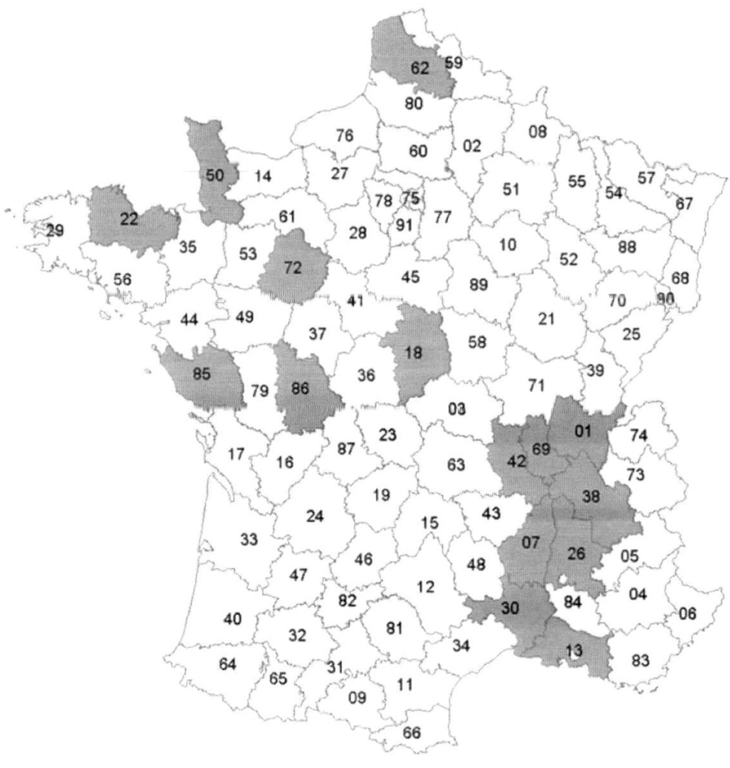

Fig. 4 Location of French departments. In *grey* are highlighted French departments in which *Dermanyssus* individuals have been found during field sampling in the present study (2005–2007)

Table 6 List of DNA-tested mite populations, including accession numbers

Pop° code	Species of mite	18S-28S rRNA	16S rRNA	COI	Country (State, French 'department')	Restricted study area	Sampling year (restricted study areas)	Host	Context
8002	*D. gallinae*			FM208713	France, 26			*Gallus gallus* (Galliform)	Layer farm
8003	*D. gallinae*			FM208733	France, 38			*Gallus gallus* (Galliform)	Layer farm
8004	*D. gallinae*			FM208722	France, 69			*Gallus gallus* (Galliform)	Little amateur hen house
8005	*D. gallinae*			FM208737				*Gallus gallus* (Galliform)	Little layer and chicken farm
8006	*D. gallinae*			FM208725 and FM208732	France, 01			*Gallus gallus* (Galliform)	Layer farm
8007	*D. gallinae*			FM208717	Belgium			*Gallus gallus* (Galliform)	Layer farm
8008	*D. gallinae*			FM208712	France, 69			*Columba livia* (Columbiform)	Breeding facilities for bird competitions
8009	*D. gallinae*			FM208724	France, 69			*Gallus gallus* (Galliform)	Layer farm
8012	*D. gallinae*			FM208739	France, 01	CB	2008 (summer)	*Gallus gallus* (Galliform)	Cages with chickens
8010	*D. gallinae*			FM881897	France, 26			*Gallus gallus* (Galliform)	Layer farm
ACA	*A. casalis*	AM903317	AM921907	AM921868	France, 69			-	Breeding facility
ADhirs	*D. hirsutus*	AM931077	AM921912	AM921878	USA, MI			*Colaptes cafer* (Piciform)	On bird
ADhirun	*D. hirundinis*	AM931076	AM921913	AM921881	USA, MI	BMOC	?	*Tachycineta bicolor* (Passeriform)	On bird
ADqui	*D. quintus*		AM931075	AM921882	USA, MI			*Picoides villosus* (Piciform)	On bird
CANIM	*D. gallinae*			FM208734	France, 69			(Pet shop with various bird species very close to each other)	Just caught from a cage with canaries and quails; many other bird species in cages next to them (including Psittaciforms, …)
CANIT	*D. gallinae*	AM903308	AM921909	AM921877	Italy			*Serinus canarius* (Fringillidae: Passeriform)	Breeding facility

Table 6 continued

Pop° code	Species of mite	18S-28S rRNA	16S rRNA	COI	Country (State, French 'department')	Restricted study area	Sampling year (restricted study areas)	Host	Context
CB3	*D. hirundinis*			FM208726	France, 01	CB	2008 (summer)	*Hirundo rustica* (Passeriform)	Nest in farm building
CB4	*D. hirundinis*			FM208727	France, 01	CB	2008 (summer)	*Hirundo rustica* (Passeriform)	Nest in farm building
CB5	*D. hirundinis*			FM208728	France, 01	CB	2008 (summer)	*Hirundo rustica* (Passeriform)	Nest in farm building
Chab	*D. gallinae*	AM931074	AM921836	AM921857	France, 01			*Gallus gallus* (Galliform)	Layer farm
CHOV	*D. hirundinis*	AM943019		FM179369	France, 72			*Hirundo rustica* (Passeriform)	Nest in a barn
COL*	*D. gallinae*	AM903307	AM921892	AM921875	France, 69			*Columbus livia* (Columbiform)	On adult bird
DR	*D. gallinae*	AM931073	AM921835		Spain			*Fringillidae* (Passeriform)	Cage
Ecop1	*D. carpathicus*			FM208731	France, 42	Ecop	2008 (winter)	*Parus* sp. (Passeriform)	Nest box
Ecop3	*D. carpathicus*			FM208729	France, 42	Ecop	2008 (winter)	*Parus* sp. (Passeriform)	Nest box
Ecop06-5	*D. carpathicus*	AM903314	AM921901	AM921873	France, 42	Ecop	2006 (summer)	*Parus major* (Passeriform)	Nest box
Ecop06-9	*D. carpathicus*			FM208730	France, 42	Ecop	2007 (summer)	*Parus caeruleus*	Nest box
ENVL07-07	*D. longipes*				France, 69	ENVL	2007 (summer)	*Parus* sp. (Passeriform)	Nest box
ENVL08-1	*D. longipes*			FM208744	France, 69	ENVL	2008 (summer)		
ENVL08-3	*D. longipes*	FM179377	FM179374	FM179365	France, 69	ENVL	2008 (summer)	*Parus* sp. (Passeriform)	Nest box
ENVL08-7	*D. longipes*					ENVL	2008 (summer)	*Parus* sp. (Passeriform)	Nest box
ENVL08-8	*D. longipes*			FM208743	France, 69	ENVL	2008 (summer)	*Parus* sp. (Passeriform)	Nest box
F01-5013	*D. gallinae*			FM208721	France, 01			*Gallus gallus* (Galliform)	Layer farm (free range)
F22AR	*D. gallinae*			FM208720	France, 22			*Gallus gallus* (Galliform)	Layer farm
F50S	*D. gallinae*			FM208719	France, 50			*Gallus gallus* (Galliform)	Layer farm
F86	*D. gallinae*			FM208718	France, 86			*Gallus gallus* (Galliform)	Layer farm
Fa1	*D. gallinae*	AM931072	AM921884	AM921853	Norway			*Gallus gallus* (Galliform)	Layer farm
Fa2	*D. gallinae*	AM931071	AM921883	AM921852	Norway			*Gallus gallus* (Galliform)	Layer farm (organic)

Table 6 continued

Pop° code	Species of mite	18S-28S rRNA	16S rRNA	COI	Country (State, French 'department')	Restricted study area	Sampling year (restricted study areas)	Host	Context
GO1	*D. apodis*	AM903299	AM921894		France, 30			*Apus apus* (Apodiform)	Nest
GO10	*D. apodis*		AM921895		France, 30			*Apus apus* (Apodiform)	Nest
GO12	*D. apodis*	AM903309			France, 30			*Apus apus* (Apodiform)	Nest
GO15	*D. apodis*	AM903313	AM921896		France, 30			*Apus apus* (Apodiform)	On young birds at nest
GO16	*D. apodis*	AM903313			France, 30			*Apus apus* (Apodiform)	On young birds at nest
GO26	*D. apodis*		AM921900		France, 30			*Apus apus* (Apodiform)	Nest
GO36	*D. apodis*			FM179371	France, 30			*Apus apus* (Apodiform)	Nest
GO44	*D. apodis*		AM921898		France, 30			*Apus apus* (Apodiform)	Nest
GO46	*D. apodis*		AM921897		France, 30			*Apus apus* (Apodiform)	Nest
GO54	*D. apodis*	AM930888		AM921874	France, 30			*Apus apus* (Apodiform)	Nest
GO58a	*D. apodis*			FM179370	France, 30			*Apus apus* (Apodiform)	Nest
GO8*	*D. gallinae*		AM921893		France, 30			*Apus apus* (Apodiform)	Nest
HIR1	*D. hirundinis*	FM179379		FM179366	France, 85	HIR	2008 (winter)	*Hirundo rustica* (Passeriform)	Nest
HIR2	*D. hirundinis*				France, 85	HIR	2008 (winter)	*Hirundo rustica* (Passeriform)	Nest
HIR5	*D. hirundinis*				France, 85	HIR	2008 (spring)	*Hirundo rustica* (Passeriform)	Nest
HIR6	*D. hirundinis*			FM208741 and FM208740	France, 85	HIR	2008 (spring)	*Hirundo rustica* (Passeriform)	Nest
HR	*D. hirundinis*	AM903300	AM921888	AM921872	France, 69			*Hirundo rustica* (Passeriform)	Nest in farm building
IL302	*D. gallinae*			FM207495	The Netherlands	IL	2007	*Sturnus vulgaris* (Passeriform)	Nest box
IL213	*D. gallinae*	FM207490	FM207492	FM207499	The Netherlands	IL	2007	*Sturnus vulgaris* (Passeriform)	Nest box
IL227	*D. gallinae*	FM207491	FM207494	FM207496	The Netherlands	IL	2007	*Sturnus vulgaris* (Passeriform)	Nest box

Table 6 continued

Pop° code	Species of mite	18S-28S rRNA	16S rRNA	COI	Country (State, French 'department')	Restricted study area	Sampling year (restricted study areas)	Host	Context
IL202	*D. gallinae*			FM207497 and FM207498	The Netherlands	IL	2007	*Sturnus vulgaris* (Passeriform)	Nest box
JBO27	*D. gallinae*			FM208716	France, 13			*Parus major* (Passeriform)	Nest box in an apple orchard
JBO46	*D. gallinae*			FM208736	France, 13			*Parus major* (Passeriform)	Nest box in an apple orchard
JBO517	*D. gallinae*	AM930885		AM921879	France, 13			*Parus major* (Passeriform)	Nest box in an apple orchard
JBO59	*D. carpathicus*	AM930882	AM921902	AM921870	France, 13			*Parus major* (Passeriform)	Nest box in an apple orchard
JMC10	*D. carpathicus*	AM943018		AM943021	France, 62			*Parus major* (Passeriform)	Nest
LB074	*D. gallinae*			AM921866	France, 18	LB	2006-2007	*Delichon urbica* (Passeriform)	Nest
LB18	*D. gallinae*	AM930889	AM921908	AM921867	France, 18	LB	2006-2007	*Delichon urbica* (Passeriform)	Nest
LC	*D. gallinae*	AM903306	AM921891	AM921859	France, 26			*Columbus livia* (Columbiform)	Breeding facility, rural country
LC23	*D. carpathicus*			FM208735 and FM881898	France, 26			*Parus major* (Passeriform)	Nest
LC10A	*D. carpathicus*			FM179367				*Parus major* (Passeriform)	Nest
LR20A	*D. carpathicus*			FM179368				*Parus* sp. (Passeriform)	Nest
MAR	*D. apodis*	AM945880	AM921899	AM921880	France, 69			*Apus apus* (Apodiform)	On young bird fallen from nest
Ob	*O. bacoti*	AM903318	AM921905	FM179677	?			*rodents*	From a lab strain in MNHN (O. Bain, Lab of Parasitology)
OC	*D. hirundinis*	AM903312	AM921889	AM921862	France, 38			*Delichon urbica* (Passeriform)	On young birds at nest

Table 6 continued

Pop° code	Species of mite	18S-28S rRNA	16S rRNA	COI	Country (State, French 'department')	Restricted study area	Sampling year (restricted study areas)	Host	Context
PARATR	*D. hirundinis*			FM208746	USA, MI	BMOC	2007 (autumn)	*Parus atricapillus* (Passeriform)	Nest box ?
Parm	*D. carpathicus*	AM903315			France, 69			*Parus major* (Passeriform)	Nest box
PAS	*D. longipes*	AM903310	AM921904	AM921869	France, 13			*Passer montanus* (Passeriform)	Nest
Percnobis	*D. gallinae*	AM943020		FM208738	France, 07			*Neophron percnopterus* (Ciconiiform)	Nest
PI*	*D. gallinae*	FM179378	FM179375	AM921860	France, 13			*Columbus livia* (Columbiform)	From nest inside a flat in city center
PO1	*D. gallinae*	AM903302		AM921854	Poland			*Gallus gallus* (Galliform)	Layer farm
PO2	*D. gallinae*		AM921914	AM921855	Poland			*Gallus gallus* (Galliform)	Layer farm
ROL1	*D. gallinae*	AM903304	AM921910	AM921864	France, 13			*Coracias garrulus* (Coraciiform)	On adult birds
ROL2	*D. gallinae*	AM903305	AM921911	AM921865	France, 13			*Coracias garrulus* (Coraciiform)	On young birds at nest
RQ	*D. carpathicus*	AM903316	AM921903	AM921876	France, 42	RQ	2006 (summer)	*Phoenicurus phoenicurus* (Passeriform)	Nest, near a house at altitude of ca. 1000 m
RQ-Mes	*D. carpathicus*			FM208715, FM208714 and FM208723	France, 42	RQ	2006 (summer)	*Parus* sp. (Passeriform)	Nest, in the wall of a house, alt ca. 1000 m
SB	*D. gallinae*			AM921858	France, 69			*Gallus gallus* (Galliform)	Small, amateur hen house
SK	*D. gallinae*	AM903303	AM921887	AM921856	Denmark			*Gallus gallus* (Galliform)	Layer farm
TACBIC	*D. hirundinis*			FM208745	USA, MI	BMOC	2007 (autumn)	*Tachycineta bicolor* (Passeriform)	Nest box
TROAED	*D. hirundinis*			FM208747	USA, MI	BMOC	2007 (autumn)	*Troglodytes aedon* (Passeriform)	Nest box

Table 6 continued

Pop° code	Species of mite	18S-28S rRNA	16S rRNA	COI	Country (State, French 'department')	Restricted study area	Sampling year (restricted study areas)	Host	Context
TPYR	*T. pyri*	FM179376	FM179373	FM179364	France			–	From a lab strain in Supagro, Montpellier (S. Kreiter)
Veol	*D. carpathicus*			AM921871	France, 69			*Parus major* (Passeriform)	Nest box
Woodp	*D. gallinae*	AM903301	AM921890	AM921863	France, 69			*Dendrocopos major* (Piciform)	On wild adult female bird

Each population corresponds to a group of *Dermanyssus* mites from a single nest (or a single building in a farm or breeding facility). There was never more than 1 species of *Dermanyssus* per nest

References

Brännström S, Morrison DA, Mattsson JG, Chirico J (2008) Genetic differences in internal transcribed spacer 1 between *Dermanyssus gallinae* from wild birds and domestic chickens. Med Vet Entomol 22(2):152–155. doi:10.1111/j.1365-2915.2008.00722.x

Carroll H, Beckstead W, OConnor T, Ebbert M, Clement M, Snell Q, McClellan D (2007) DNA reference alignment benchmarks based on tertiary structure of encoded proteins. Bioinformatics 23(19):2648–2649. doi:10.1093/bioinformatics/btm389

De Lillo E (2001) A modified method for Eriophyoid mite extraction (Acari: Eriophyoidea). Int J Acarol 27(1):67–70

De Souza Amorim D (1982) Classificaçao por sequenciaçao: uma proposa para a denominaçao dos ramos retardados. Rev Bras Zool 11:1–9

Flechtmann HW, Baggio D (1993) On Phoresy of hematophagous ectoparasitic acari (Parasitiformes: Ixodidae and Dermanyssidae) on Coleoptera observed in Brazil. Int J Acarol 19(2):195–196

Hoogstraal H, Aeschlimann A (1982) Tick host specificity. Mitteilungen des Schweizerischen entomologischen Gesellschaft 55:5–32

Huelsenbeck JP, Ronquist F (2001) MRBAYES: Bayesian inference of phylogeny. Bioinformatics 17: 754–755. doi:10.1093/bioinformatics/17.8.754

Knee W (2008) Five new species of Rhinonyssidae (Mesostigmata) and one new species of *Dermanyssus* (Mesostigmata: Dermanyssidae) from birds of Alberta and Manitoba, Canada. J Parasitol 94:348–374. doi:10.1645/GE-1358.1

Lesna I, Wolfs P, Faraji F, Roy L, Komdeur J, Maurice W. Sabelis (2009) Candidate predators for biological control of the poultry red mite *Dermanyssus gallinae*. Exp Appl Acarol. doi:10.1007/s10493-009-9239-1

Livezey BC, Zusi RL (2007) Higher order phylogeny of modern birds (Theropoda, Aves: Neornithes) based on comparative anatomy II. Analysis and discussion. Zool J Linn Soc 149:1–95. doi:10.1111/j.1096-3642.2006.00293.x

Moss WW (1978) The mite genus *Dermanyssus*: a survey, with description of *Dermanyssus trochilinis*, n. sp., and a revised key to the species (Acari: Mesostigmata: Dermanyssidae). J Med Ent 14(6):627–640

Nylander JAA (2004) MrModeltest v2. Program distributed by the author. Evolutionary Biology Centre, Uppsala University

Peterson AP (2008) http://www.zoonomen.net/avtax/frame.html, Version 7.07 (2008.07.30)

Ronquist F, Huelsenbeck JP (2003) MRBAYES 3: Bayesian phylogenetic inference under mixed models. Bioinformatics 19:1572–1574. doi:10.1093/bioinformatics/btg180

Roy L, Chauve CM (2007) Historical review of the genus *Dermanyssus* Dugès, 1834 (Acari: Mesostigmata: Dermanyssidae). Parasite 14(2):87–100

Roy L, Dowling APG, Chauve CM, Buronfosse T (2008) Delimiting species boundaries within *Dermanyssus* Dugès, 1834 (Acari: Mesostigmata) using a total evidence approach. Mol Phylogenet Evol. doi:10.1016/j.ympev.2008.11.012

Swofford DL (2001) PAUP*–Phylogenetic Analysis Using Parsimony (*and other methods). Version 4.0B 10. [Computer software and manual]. Sinauer Associates, Sunderland, Massachusetts

Valera F, Casas-Crivillé A, Hoi H (2003) Interspecific Parasite exchange in a mixed colony of birds. J Parasitol 89(2):245–250

Valiente Moro C, Chauve C, Zenner L (2005) Vectorial role of some dermanyssoid mites (Acari, Mesostigmata, Dermanyssoidea). Parasite 12(2):99–109

Valiente Moro C, Fravalo P, Amelot M, Chauve C, Zenner L, Salvat G (2007) Colonization and invasion in chicks experimentally infected with *Dermanyssus gallinae* contaminated by *Salmonella enteritidis*. Avian Pathol 36(4):307–311. doi:10.1080/03079450701460484

Wilm A, Mainz I, Steger G (2006) An enhanced RNA alignment benchmark for sequence alignment programs. Algorithms Mol Biol 1(19):1–11

Phylogenetic relationship between *Dermanyssus gallinae* populations in European countries based on mitochondrial COI gene sequences

M. Marangi · C. J. de Luna · M. A. Cafiero · A. Camarda · S. le Bouquin · D. Huonnic · A. Giangaspero · O. A. E. Sparagano

Originally published in the journal Experimental and Applied Acarology, Volume 48, Nos 1–2, 143–155.
DOI: 10.1007/s10493-009-9237-3

Abstract Phylogenetic analysis of *Dermanyssus gallinae* mites originating from UK, France and Italy was performed using partial amplification of the mitochondrial COI gene. Results showed that UK samples reveal the greatest variation and diversity and are linked to one of the French populations highlighting North–South genetic transitions in European red mite populations. Intra-farm variations between mite samples highlighted the diversity between national populations and possibly its origin from the different chemical strategies used in each country.

Keywords Phylogeny · *Dermanyssus gallinae* · Population diversity · Mitochondrial COI

Introduction

The blood-feeding mite *Dermanyssus gallinae* (De Geer), also known as the poultry red mite or chicken mite, has worldwide distribution and is an economically important parasitic pest of domestic chickens (Chauve 1998). Affected birds show decreased production,

M. Marangi · A. Giangaspero (✉)
Dipartimento PrIME, Centro Interdipartimentale BIOAGROMED, Università degli Studi di Foggia, 71100 Foggia, Italy
e-mail: a.giangaspero@unifg.it

C. J. de Luna · O. A. E. Sparagano
School of Agriculture, Food and Rural Development, Newcastle University, Newcastle upon Tyne NE1 7RU, UK

M. A. Cafiero
Istituto Zooprofilattico Sperimentale della Puglia e Basilicata, Foggia, Italy

A. Camarda
Dipartimento di Sanità pubblica e Zootecnia, Università di Bari, Bari, Italy

S. le Bouquin · D. Huonnic
Unité EBEAC, AFSSA-BP 53, 22 440 Ploufragan, France

irritation and, in severe infection, anaemia leading to death. The mite spends most of its life aggregated in cracks and crevices in the poultry house, where mating, reproduction, and development take place. The parasite only visits its host for blood meals about every 3 day (Kirkwood 1963), and may, under optimal conditions in poultry production systems, complete the feeding-oviposition cycle within 1 week, resulting in the rapid establishment of dense populations of mites.

Dermanyssus gallinae has been considered as a potential vector of pathogens causing fowl spirochaetosis, chicken pox, encephalitis, fowl cholera and Newcastle disease (Zeman et al. 1982; Chauve 1998). *Erysipelothrix rhusiopathiae*, a bacteria causing septicaemia in hens, has been isolated from mites originating from poultry facilities with outbreaks of erysipelas (Chirico et al. 2003). Recently, other pathogens were discovered to be present in *D. gallinae* (Valiente Moro et al. 2008) and the vector capacity for salmonellosis has been demonstrated (Valiente Moro et al. 2007). The zoonotic potential of these pathogens should also be considered as additional concerns associated with the parasite, as *D. gallinae*, in the absence of its definitive host, occasionally attacks dogs, cats, rodents, horses and humans (Brockis 1980), causing dermatitis and intense pruritus.

Dermanyssus gallinae is in large expansion from north to south parts of Europe. Although prevalence studies seem to prove the endemicity of this poultry pest (Sparagano et al. 2008) other studies are showing a diverse range of acaricide resistance (Marangi et al. 2008) highlighting the possibility of high variability within *Dermanyssus* populations in different countries.

To date, no molecular studies have assessed genetic variability patterns among worldwide populations of *D. gallinae*. Molecular data have long been recognized as an attractive marker for phylogenetic analysis, for instance, allozyme polymorphism (Puterka et al. 1993; Sperling et al. 1997), RFLP (Kambhampati and Rai 1991) and randomly amplified polymorphic DNA (RAPD) (Puterka et al. 1993). Recently, DNA sequences of various genomic regions such as the mitochondrial cytochrome oxidase I (COI) and the ribosomal internal transcribed spacers (ITS1 and ITS2) have been used to infer phylogenies in a wide range of organisms (Hills and Dixon 1991; Simon et al. 1994; Brannstrom et al. 2008). The study of phylogeny is important for understanding species diversity, phylogenetic patterns, and evolutionary processes. In pest species, this is also important in the development of biological pest control strategies (Cruickshank 2002).

In this study, we analyzed DNA sequences of *D. gallinae* obtained from three different European countries: Italy, France and UK, using partial sequences of COI to evaluate the genetic relationship between them. We used a genetic tree based on the mitochondrial DNA to assess whether the mites showed diverged lineages. We have targeted the COI gene because it is highly polymorphic within species, and is therefore suitable for analyses of intra specific variations (Toda et al. 2000; Ballard and Rand 2005).

Materials and methods

Mite samples

Mite samples were collected from different caged laying hen poultry farms in Europe. In each farms, the mites were collected from different hiding sites i.e., under feed troughs, within fittings and fastening clips of cage support, under egg conveyor belt, under manure belt.

One farm (UK farm) was from the North-East of UK, two farms from the French Brittany region (F1 and F2 farms) and three farms from the Italian Apulia region (I, I2 and

13 farms). In the investigated farms, frequent treatments with acaricides were applied, as stated by the farmers, and organophosphates (OPs), amidine and pyrethroid-based acaricides were the most commonly used. Mites were morphologically identified to the species level as *D. gallinae*. Individuals were kept directly frozen at −20°C. For each farm, 30 unfed adult mites were tested.

DNA extraction and detection of *Dermanyssus gallinae*

DNA was extracted from individual mites using Qiamp DNA extraction kit (Qiagen, Hilden, Germany). DNA extraction was performed according to the manufacturer's tissue protocol, with some adjustments due to the weight of one *D. gallinae*, compared to the recommended tissue weight of 25 mg: briefly, mites were crushed in 0.45 μl PBS and volume was adjusted with 1.81 μl ATL buffer. Other reactive volumes were also modified (0.18 μl proteinase K, 1.81 μl AL buffer, 1.81 μl ethanol, 4.72 μl AW1 and AW2 buffers). The digestion step with proteinase K was prolonged overnight at 56°C. The genetic material was finally eluted with 30 μl of AE buffer and stored at −20°C.

The efficiency of the DNA extraction method was evaluated by amplifying the mitochondrial 16S rRNA gene of *D. gallinae*, using specific primers designed for its 16S rDNA sequence (accession number L34326), F16 (5′-TGGGTGCTAAGAGAATGGATG-3′) and R16 (5′-CCGGTCTGAACTCAGATCAAG-3′), which amplify a 377 bp region (Desloire et al. 2006). PCR reactions were performed in 25 μl volume containing 10 μm of each oligonucleotide primer, 12.5 μl of REDTaq ReadyMix PCR Reaction Mix (Sigma–Aldrich) and 2 μl of DNA extract. PCR amplifications were performed in an eppendorf mastercycler gradient, with a cycling program consisting of a 10 min denaturation step at 94°C followed by 39 cycles of denaturation (1 min, 94°C), annealing (45 s, 55°C), and extension (1 min, 72°C), and a final extension step of 10 min at 72°C. PCR products were separated by a 1.2% agarose gel electrophoresis containing ethidium bromide, and DNA fragments were visualized with a multigenius bioImaging system (Syngene). After a first amplification, each negative DNA sample was submitted to a second PCR amplification, and the results were recorded.

PCR amplification and DNA sequencing

For the phylogenetic analysis the central part of the mitochondrial COI gene was amplified using primers COIF (5′-TGATTTTTTGGTCACCCAGAAG-3′) and COIR (5′-TAC AGCTCCTATAGATAAAAC-3′) (Navajas et al. 1994; Ros and Breeuwer 2007) which produced a fragment of about 400 bp. PCR was performed in a 25 μl volume containing 10 μm of each oligonucleotide primer, 12.5 μl of REDTaq ReadyMix PCR Reaction Mix (Sigma–Aldrich), 1.25 μl of BSA and 5 μl of DNA extract. PCR amplifications were performed in an eppendorf mastercycler gradient, with a cycling program consisting of a 4 min denaturation step at 94°C followed by 44 cycles of denaturation (1 min, 94°C), annealing (1 min, 53°C), and extension (1 min, 72°C), and a final extension step of 4 min at 72°C. PCR products were separated by a 1.2% agarose gel electrophoresis containing ethidium bromide, and DNA fragments were visualized with the multigenius bioimaging system (Syngene).

PCR products were purified using a PCR Montage KIT (Millipore) and were directly sequenced using the ABI PRIMS BygDye Terminator sequences KIT v. 3.1 (Applied Biosystem, Nieuwerkerk a/d IJssel, The Netherlands) according to the manufacturer's instructions. Sequences were run on an ABI 3130 automated DNA sequencer. Double stranded sequences were obtained for all samples, for accuracy. Nucleotide sequences were aligned using the sequence alignment program Clustal X (Thompson et al. 1997) and compared to the sequences obtained from the GenBank database. Sequences that differed by one or more nucleotides were considered as different haplotypes, while sequences exhibiting identical SNPs at same nucleotide positions were considered as similar haplotypes. Mitochondrial DNA COI partial amino acid sequence alignment was also carried out using the sequence alignment program Clustal X.

Phylogenetic analysis

Phylogenetic analyses were conducted using PHYLIP (Phylogeny Inference Package) version 3.5c. This software was used to calculate numbers of variable sites, uncorrected genetic distance, nucleotides composition, and transition and transversion ratios (Sullivan et al. 2005; Swofford 2002; Swofford and Sullivan 2003).

The genetic distance, uncorrected *P*, was calculated for each pair of sequences, and then transitional and transversional substitution where compared with these genetic distances.

The phylogenetic tree was constructed based on neighbour-joining (NJ) algorithms (*P*-distance) using PHYLIP. For the NJ analyses robustness of nodes was assessed with 1,000 NJ-bootstrap replicates. Phylogenetic analysis was performed for: (1) the sequences of each farm and (2) the sequences of all the farms.

Results

Sequences analysis

Thirty-four good quality DNA sequences of*D. gallinae* were analyzed ($n = 4$ from UK farm, F1 and I farms, $n = 6$ from F2 farm, $n = 7$ from I2 farm and $n = 9$ from I3 farm).

A 365 bp fragment of the COI gene was obtained from all the sequences. Figure 1 shows the alignment of nucleotides sequences. One hundred-eighty eight (51.5%) sites were conservative and 177 (48.5%) were phylogenetically informative sites. Within conservative sites the nucleotides composition was 40.6% T, 31.6% A, 14.8% C and 13% G. This high AT content (40.6 + 31.6%) is a general feature of the COI mitochondrial DNA region in arthropods, and is similar to other studies on insects and mites (Simon et al. 1994; Mangold et al. 1998).

The average nucleotide transition per site was 0.7 and the transversion rate was 1.07. The rate of transversions (61.2%) was predominant. Transversion frequencies were 36.6% for A–T, 21% for A–C, 24% for T–G and 9% for C–G. Transition frequencies were 51.2% for C–T and 42.1% for A–G (Hinomoto et al. 2001; Hinomoto and Takafuji 2001).

The pairwise uncorrected genetic distance (*P*) calculated from all the sequences ranged from 0.0027 to 0.3014.

Intra genetic variability and pairwise comparisons between samples

UK samples

The alignment of four UK COI sequences reveals that there were 66 (18%) nucleotides variations. In 57.6% of these variations, three UK samples show the same nucleotide change while one sample (9UK) shows a different nucleotide. In particular this sample

Fig. 1 Sequences of mitochondrial COI DNA from 34 strains of *Dermanyssus gallinae*

```
  9I2   ATTACCCA-T ATTGTTT--- GTTATCAAA- -CTGG-AAAA AAGAAA-CCT
 11I2   ATTACCCA-T ATTGTTT--- GTTATCAAA- -CTGG-AAAA AAGAAA-CCT
 10F1   ATTACCCC-A TTAGTTT--- GTTACCAAA- -CTGG-AAAA AAGTAA-CCT
  2F1   ATTACCCC-A TTCGTTT--- GTTAGCAAA- -CTGG-AAAA AAGAAA-CCT
 15F1   ATTACCCC-A GTCGTTT--- GTTACCAAA- -CCGG-AAAA AAGAAA-CCT
 18F1   ATTACCCC-A GTAGTTT--- GTTATCAAA- -CTGG-AAAA AAGAAA-CCT
  5I3   ATTTCCCA-T ATTGTTT--- GTTATCAAA- -CTGG-AAAA AAGAAA-CCT
  2I3   ATTTCCCA-T ATTGTTT--- GTTATCAAA- -CTGG-AAAA AAGAAA-CCT
  1I3   ATTTCCCA-T ATTGTTT--- GTTATCAAA- -CTGG-AAAA AAGAAA-CCT
 10I3   ATTTCCCA-T ATTGTTT--- GTTATCAAA- -CTGG-AAAA AAGAAA-CCT
  7I3   ATTTCCCA-T ATTGTTT--- ATTATCAAA- -CTGG-AAAA AAGAAA-CCT
 12I3   ATTTCCCA-T ATTGTTT--- GTTATCAAA- -CTGG-AAAA AAGAAA-CCT
  8I3   ATTTCCCA-T ATTGTTT--- GTTATCAAA- -CTGG-AAAA AAGAAA-CCT
  6I3   ATTTCCCA-T ATTGTTT--- GTTATCAAA- -CTGG-AAAA AAGAAA-CCT
 16I3   ATTTCCCA-T ATTGTTT--- GTTATCAAA- -CTGG-AAAA AAGAAA-CCT
  1I2   ATTACCCA-T ATTGTTT--- GTTATCAAA- -CTGG-AAAA AAGAAA-CCT
 17I2   ATTACCCA-T ATTGTTT--- GTTATCAAA- -CTGG-AAAA AAGAAA-CCT
  5I2   ATTACCCA-T ATTGTTT--- GTTATCAAA- -CTGG-AAAA AAGAAA-CCT
  6I2   ATTACCCA-T ATTGTTT--- GTTATCAAA- -CTGG-AAAA AAGAAA-CCT
 18I    ATTACCAT-T ATGGGTT--- GTTATCAAC- -CTGG-AAAA AAGAAA-CCT
 13I    ATTTCCAT-T ATTGGTC--- GATATCAAC- -CTGG-AAAA AAGAAA-CCC
 20I    ATTTCCAT-T GTTTGTA--- GATATGAAC- -CTGG-AAAA AAGAAA-CCT
 12I    ATTACCAT-T ATTTGTT--- GGTATCAAC- -CTGA-AAAA AAGAAA-CCC
  7I2   ATTTCCCA-T ATTGTTT--- GTTATCAAA- -CTGG-AAAA AAGGAA-CCT
  7F2   ATTTCCCA-T ATTGTTT--- GTTATCAAA- -CTGG-AAAA AAGAAA-CCC
  5F2   ATTTCCCA-T ATTGTTT--- GTTATCAAA- -CTGG-AAAA AAGAAA-CCT
 12F2   ATTTCCCA-T ATTGTTT--- GTTATCAAA- -CTGG-AAAA AAGAAA-CCC
  6F2   ATTTCCCA-T ATTGTTT--- GTTATCAAA- -CTGG-AAAA AAGAAA-CCT
  3F2   ATTTCCCA-T ATTGTTT--- GTTATCAAA- -CTGG-AAAA AAGAAA-CCT
  1F2   ATTTCCCA T ATTGTTT--- GTTATCAAA- -CTGG-AAAA AAGAAA-CCT
  1UK   ATTTCTCA-T ATCGTTT--- GTTACTAAA- -CGGT-GAAA AAAATA-CCT
  8UK   ATTTCTCA-T ATCGTTT--- GTTACTAAA- -CGGT-GAAA AAAATA-CCT
  9UK   ATTTCTCA-T ATCGTTT--- GTTACTAAA- -CGGT-GAAA AAAATA-CCT
 10UK   ATTTCTCA-T ATCGTTT--- GTTACTAAA- -CGGT-GAAA AAAATA-CCT

  9I2   TGTTGGAAAT ATTAGAA-TA ATCTATGCGA TATTAACAAT -TGGTATTTT
 11I2   TTTTGGAAAT ATTAGAA-TA ATCTATGCGA TATTAACAAT -TGGTATTTT
 10F1   TTTGGGTAAT ATTAGAA-TA ATCTATGCGA TATTAACAAT -TGGTATTTT
  2F1   TTTGGGAAAT ATTAGAA-TA ATCTATGCGA TATTAACAAT -TGGTATTTT
 15F1   TTTGGGAAAT ATTAGAA-TA ATCTATGCGA TATTAACAAT -CGGTATTTT
 18F1   TTTGGGATAT ATTAGAA-TA ATCTATGCGA TATTAACAAT -TCGTATTTT
  5I3   TTTTGGAAAT ATTAGAA-TA ATCTATGCGA TATTAACAAT -TGGTATTTT
  2I3   TTTTGGAAAT ATTAGAA-TA ATCTATGCGA TATTAACAAT -TGGTATTTT
  1I3   TTTTGGAAAT ATTAGAA-TA ATCTATGCGA TATTAACAAT -TGGTATTTT
 10I3   TTTTGGAAAT ATTAGAA-TA ATCTATGCGA TATTAACAAT -TGGTATTTT
  7I3   TTTTGGAAAT ATTAGAA-TA ATCTATGCGA TATTAACAAT -TGGTATTTT
 12I3   TTTTGGAAAT ATTAGAA-TA ATCTATGCGA TATTAACAAT -TGGTATTTT
  8I3   TTTTGGAAAT ATTAGAA-TA ATCTATGCGA TATTAACAAT -TGGTATTTT
  6I3   TTTTGGAAAT ATTAGAA-TA ATCTATGCGA TATTAACAAT -TGGTATTTT
 16I3   TTTTGGAAAT ATTAGAA-TA ATCTATGCGA TATTAACAAT -TGGTATTTT
  1I2   TGTTGGAAAT ATTAGAA-TA ATCTATGCGA TATTAACAAT -TGGTATTTT
 17I2   TGTTGGAAAT ATTAGAA-TA ATCTATGCGA TATTAACAAT -TGGTATTTT
  5I2   TTTTGGAAAT ATTAGAA-TA ATCTATGCGA TATTAACAAT -TGGTATTTT
  6I2   TTTTGGAAAT ATTAGAA-TA ATCTATGCGA TATTAACAAT -TGGTATTTT
 18I    TTTTGGAAAT ATTAGAA-TA ATCTATGCGA TATTAACAAT -TGGTATTTT
 13I    TTTTGGAAAT ATTAGAA-TA ATCTATGCGA TATTAACAAT -TGGTATTTT
 20I    TTTTGGAAAT ATTAGAA-TA ATCTATGCGA TATTAACAAT -TGGTATTTT
 12I    TTTTGAAAAT ATTAGAA-TA ATCTATGCGA TATTAACAAT -TGGTATTTT
  7I2   TTATGAGAGA GTTAGAA-TA ATCTATGCGA TATTAACATT -TCGTATTTT
  7F2   TTTTGGAAAT ATTAGAA-TA ATCTATGCGA TATTAACAAT -TGGTATTTT
  5F2   TTTTGGAAAT ATTAGAA-TA ATCTATGCGA TATTAACAAT -TGGTATTTT
 12F2   TTTTGGAAAT ATTAGAA-TA ATCTATGCGA TATTACTAAT -TGGTATTTT
  6F2   TTTTGGAAAT ATTAGAA-TA ATCTATGCGA TATTAACAAT -TGGTATTTT
  3F2   TTTTGGAAAT ATTAGAA-TA ATCTATGCGA TATTAACAAT -TGGTATTTT
  1F2   TTTTGGAAAT ATTAGAA-TA ATCTATGCGA TATTAACAAT -TGGTATTTT
  1UK   TTTTGGAAAT ATTAGAA-TA ATTTATGCCA AATTAACAAT -TGGTATTTT
  8UK   TTTTGGAAAT ATTAGAA-TA ATTTATGCCA AATTAACAAT -TGGTATTTT
  9UK   TTTTGGAAAT ATTAGAA-TA ATTTATGCCA AATTAACAAT -TGGTATTCT
 10UK   TTTTGAAAAT ATTAGAA-TA ATTTATGCCA AATTAACAAT -TGGTATTTT

  9I2   AGGATT-TAT TGTTTGAGCC CATCATATAT TTACAATTGG ATTAGATATT
 11I2   AGGATT-TAT TGTTTGAGCC CATCATATAT TTACAATTGG ATTAGATATT
 10F1   AGGATT-TAT TGTTTGAGCC CATCATATAT TTACAATTGG ATTAGATATT
  2F1   AGGATT-TAT TGTTTGAGCC CATCATATAT TTACAATTGG ATTAGATATT
 15F1   AGGATT-TAT TGTTTGAGCC CATCATATAT TTACAATTGG ATTAGATATT
 18F1   AGGATT-TAT TGTTTGAGCC CATCATATAT TTACAATTGG ATTAGATATT
  5I3   AGGATT-TAT TGTTTGAGCC CATCATATAT TTACAATTGG ATTAGATATT
  2I3   AGGATT-TAT TGTTTGAGCC CATCATATAT TTACAATTGG ATTAGATATT
  1I3   AGGATT-TAT TGTTTGAGCC CATCATATAT TTACAATTGG ATTAGATATT
 10I3   AGGATT-TAT TGTTTGAGCC CATCATATAT TTACAATTGG ATTAGATATT
```

Fig. 1 continued

```
  7I3   AGGATT-TAT TGTTTGAGCC CATCATATAT TTACAATTGG ATTAGATATT
 12I3   AGGATT-TAT TGTTTGAGCC CATCATATAT TTACAATTGG ATTAGATATT
  8I3   AGGATT-TAT TGTTTGAGCC CATCATATAT TTACAATTGG ATTAGATATT
  6I3   AGGATT-TAT TGTTTGAGCC CATCATATAT TTACAATTGG ATTAGATATT
 16I3   AGGATT-TAT TGTTTGAGCC CATCATATAT TTACAATTGG ATTAGATATT
  1I2   AGGATT-TAT TGTTTGAGCC CATCATATAT TTACAATTGG ATTAGATATT
 17I2   AGGATT-TAT TGTTTGAGCC CATCATATAT TTACAATTGG ATTAGATATT
  5I2   AGGATT-TAT TGTTTGAGCC CATCATATAT TTACAATTGG ATTAGATATT
  6I2   AGGATT-TAT TGTTTGAGCC CATCATATAT TTACAATTGG ATTAGATATT
 18I    AGGATT-TAT TGTTTGAGCC CATCATATAT TTACAATTGG ATTAGATATT
 13I    AGGATT-TAT TGTTTGAGCC CATCATATAT TTACAATTGG ATTAGATATT
 20I    AGGATT-TAT TGTTTGAGCC CATCATATAT TTACAATTGG ATTAGATATT
 12I    AGGATT-TTT TGTTTGAGCC CATCATATAT TTACAATTGG ATTAGATATT
  7I2   AGGATT-TAT TGTTTGAGCC CATCATATAT TTACAATTGG ATTAGATATT
 7F2    AGGATT-TAT TGTTTGAGCC CATCATATAT TTACAATTGG ATTAGATATT
 5F2    AGGATT-TAT TGTTTGAGCC CATCATATAT TTACAATTGG ATTAGATATT
12F2    AGGATT-TAT TGTGTTAACC CATCATATAT TTACAATTGG ATTAGATATT
 6F2    AGGATT-TAT TGTTTGAGCC CATCATATAT TTACAATTGG ATTAGATATT
 3F2    AGGATT-TAT TGTTTGAGCC CATCATATAT TTACAATTGG ATTAGATATT
 1F2    AGGATT-TAT TGTTTGAGCC CATCATATAT TTACAATTGG ATTAGATATT
 1UK    AGGCTT-CAT TGTTTGAGCA CATCATATAT TTACAATTGG ACTTGATATT
 8UK    AGGCTT-CAT TGTTTGAGCA CATCATATAT TTACAATTGG ACTTGATATT
 9UK    AGGTTT-CAT TGTTTGAGCA CATCATATAT TTACAATTGG ACTTGATATC
10UK    AGGCTT-CAT TGTTTGAGCA CATCATATAT TTACAATTGG ACTTGATATT

  9I2   GATACTCG-A GC---TTATT TT-ACTGCCG CCACAATGAT T-ATTGCAAT
 11I2   GATACTCG-A GC---TTATT TT-ACTGCCG CCACAATGAT T-ATTGCAAT
10F1    GATACTCG-A GC---TTATT TT-ACTGCCG CCACAATGAT T-ATTGCCAT
 2F1    GATACTCG-A GC---TTATT TT-ACTGCCG CCACAATGAT T-ATTGCAAT
15F1    GATACTCG-A GC---TTATT TT-ACTGCCG CCACAATGAT T-ATTGCAAT
18F1    GATACTCG-A GC---TTATT TT-ACTGCCG CCACAATGAT T-ATTGCAAT
  5I3   GATACTCG-A GC---TTATT TT-ACTGCCG CCACAATGAT T-ATTGCAAT
  2I3   GATACTCG-A GC---TTATT TT-ACTGCCG CCACAATGAT T-ATTGCAAT
  1I3   GATACTCG-A GC---TTATT TT-ACTGCCG CCACAATGAT T-ATTGCAAT
 10I3   GATACTCG-A GC---TTATT TT-ACTGCCG CCACAATGAT T-ATTGCAAT
  7I3   GATACTCG-A GC---TTATT TT-ACTGCCG CCACAATGAT T-ATTGCAAT
 12I3   GATACTCG-A GC---TTATT TT-ACTGCCG CCACAATGAT T-ATTGCAAT
  8I3   GATACTCG-A GC---TTATT TT-ACTGCCG CCACAATGAT T-ATTGCAAT
  6I3   GATACTCG-A GC---TTATT TT-ACTGCCG CCACAATGAT T-ATTGCAAT
 16I3   GATACTCG-A GC---TTATT TT-ACTGCCG CCACAATGAT T-ATTGCAAT
  1I2   GATACTCG-A GC---TTTTT TT-GCTGCCG CCACAATGAT T-ATTGCAAT
 17I2   GATACTCG-A GC---TTATT TT-ACTGCCG CCACAATGAT T-ATTGCAAT
  5I2   GATACTCG-A GC---TTATT TT-ACTGCCG CCACAATGAT T-ATTGCAAT
  6I2   GATACTCG-A GC---TTATT TT-ACTGCCG CCACAATGAT T-ATTGCAAT
 18I    GATACTCG-A GC---TTATT TT-ACTGCCG CCACAATGAT T-ATTGCAAT
 13I    GATACTCG-A GC---TTATT TT-ACTGCCG CCACAATGAT T-ATTGCAAT
 20I    GATACTCG-A GC---TTATT TT-ACTGCCG CCACAATGAT T-ATTGCAAT
 12I    GATACTCG-A GC---TTATT TT-ACTGCCG CCACAATGAT T-ATTGCAAT
  7I2   GATACTCG-A GC---TTATT TT-ACTGCCG CCACAATGAT T-ATTGCAAT
 7F2    GATACTCG-A GC---TTATT TT-ACTGCCG CCACAATGAT T-ATTGCAAT
 5F2    GATACTCG-A GC---TTATT TT-ACTGCCG CCACAATGAT T-ATTGCAAT
12F2    GTTACTCG-A GC---TTATT TT-ACTGCCG CCACAATGAT T-ATTGCAAT
 6F2    GATACTCG-A GC---TTATT TT-ACTGCCG CCACAATGAT T-ATTGCAAT
 3F2    GATACTCG-A GC---TTATT TT-ACTGCCG CCACAATGAT T-ATTGCAAT
 1F2    GATACTCG-A GC---TTATT TT-ACTGCCG CCACAATGAT T-ATTGCAAT
 1UK    GATACTCG-A GC---AAATT TT-ACTTCTG CCACTATAAT T-ATTGCAAT
 8UK    GATACTCG-A GC---AAATT TT-ACTTCTG CCACTATAAT T-ATTGCAAT
 9UK    GATACTCG-A GC---ATATT TT-ACTTCTG CCACTGTAAA T-ATTGCAAT
10UK    GATACTCG-A GC---AAATT TT-ACTTCTG CCACTATAAT T-ATTGCAAT

  9I2   TCCAACAGGA A-TTAAGATT TTTTCA-TGA ATTTCAACA- -CTCCACGGA
 11I2   TCCAACAGGA A-TTAAGATT TTTTCA-TGA ATTTCAACA- -CTCCACGGA
10F1    TCCAACAAGG A-TTAAGATT TTTTCA-TGA ATTTCAACC- -CTCCCCCGG
 2F1    TCCAAACAGG A-TTAAGATT TTTTCA-TGA ATTTCAACA- -CTCCACGGG
15F1    TCCAACAGGA A-TTAAGATT TTTTCA-TGA ATTACAACA- -CTCCACGGA
18F1    TCCAACAGGA A-TTAAGATT TTTTCA-TGA ATTTCAACA- -CTCCACGGA
  5I3   TCCAACAGGA A-TTAAGATT TTTTCA-TGA ATTTCAACA- -CTCCACGGA
  2I3   TCCAACAGGA A-TTAAGATT TTTTCA-TGA ATTTCAACA- -CTCCACGGA
  1I3   TCCAACAGGA A-TTAAGATT TTTTCA-TGA ATTTCAACA- -CTCCACGGA
 10I3   TCCAACAGGA A-TTAAGATT TTTTCA-TGA ATTTCAACA- -CTCCACGGA
  7I3   TCCAACAGGA A-TTAAGATT TTTTCA-TGA ATTTCAACA- -CTCCACGGA
 12I3   TCCAACAGGA A-TTAAGATT TTTTCA-TGA ATTTCAACA- -CTCCACGGA
  8I3   TCCAACAGGA A-TTAAGATT TTTTCA-TGA ATTTCAACA- -CTCCACGGA
  6I3   TCCAACAGGA A-TTAAGATT TTTTCA-TGA ATTTCAACA- -CTCCACGGA
 16I3   TCCAACAGGA A-TTAAGATT TTTTCA-TGA ATTTCAACA- -CTCCACGGA
  1I2   ACCAACAGGA A-TTAAGATT TTTTCA-TGA ATTTCAACA- -CCTCACGGA
 17I2   TCCAACAGGA A-TTAAGATT TTTTCA-TGA ATTTCAACA- -CTCCACGGA
  5I2   TCCAACAGGA A-TTAAGATT TTTTCA-TGA ATTTCAACA- -CTCCACGGA
  6I2   TCCAACAGGA A-TTAAGATT TTTTCA-TGA ATTTCAACA- -CTCCACGGA
```

Fig. 1 continued

```
18I    TCCCACAGGA A-TTAAATTT TTTTCA-TGA ATTTCAACA- -CTCCACGGA
13I    TCCCACAGGA A-TTAAATTT TTTTCA-TGA ATTTCAACA- -CTCCACGGA
20I    TCCAACAGGA A-TTAAGATT TTTTCA-TGA ATTTCAACA- -CTCCACGGA
12I    TCCAACAGGA A-TTAAGATT TTTTCA-TGA ATTTCAACA- -CTCCGCGTA
7I2    TCCAACAGGA A-TTAAGATT TTTTCA-TGA ATTTCAACA- -CTCCACGGA
7F2    TCCAACAGGA A-TTAAGATT TTTTCA-TGA ATTTCAACA- -CTCCACGGA
5F2    TCCAACAGGA A-TTAAGATT TTTTCA-TGA ATTTCAACA- -CTCCACGGA
12F2   TCCAACAGGA A-TTAAGATT TTTTCA-TGA ATTTCAACA- -CTCCACGGA
6F2    TCCAACAGGA A-TTAAGATT TTTTCA-TGA ATTTCAACA- -CTCCACGGA
3F2    TCCAACAGGA A-TTAAGATT TTTTCA-TGA ATTTCAACA- -CTCCACGGA
1F2    TCCAACAGGA A-TTAAGATT TTTTCA-TGA ATTTCAACA- -CTCCACGGA
1UK    CCCAACAGGA A-TTAAAATG TTTTCT-TGA ATTTCTACA- -TTCTACGTT
8UK    CCCAACAGGA A-TTAAAATG TTTTCT-TGA ATTTCTACA- -TTCTACGTT
9UK    CCCAACAGGA A-TTAAAATG TTTTCT-CGA ATTTATACA- -CTTCATGGT
10UK   CCCAACAGGA A-TTAAAATG TTTTCT-TGA ATTTCTACA- -TTCTACGTT

9I2    -GTAAA-TAT TAATTTTAAC CC-ACCGATT CTTTGATC-T TTAGGA----
11I2   -GTAAA-TAT TAATTTTAAC CC-ACCGATT CTTTGATC-T TTAGGA----
10F1   -GCAAA-TAT GAATTTTAAC CC-ACCGATT CTATGGTC-T TTAGGA----
2F1    -GCAAA-TAT TAATTTTAAC CC-ACCGATT CTTTGATC-T TTAGGA----
15F1   -ACAAA-TAT TAATTTTAAC CC-ACCGATT CTTTGACC-T TTAGGA----
18F1   -GCAAA-TAT TAATTTTAAC CC-ACCGATT CTTTGATC-T TTAGGA----
5I3    -GCAAA-TAT TAATTTTAAC CC-ACCGATT CTTTGATC-T TTAGGA----
2I3    -GCAAA-TAT TAATTTTAAC CC-ACCGATT CTTTGATC-T TTAGGA----
1I3    -GCAAA-TAT TAATTTTAAC CC-ACCGATT CTTTGATC-T TTAGGA----
10I3   -GCAAA-TAT TAATTTTAAC CC-ACCGATT CTTTGATC-T TTAGGA----
7I3    -GCAAA-TAT TAATTTTAAC CC-ACCGATT CTTTGATC-T TTAGGA----
12I3   -GCAAA-TAT TAATTTTAAC CC-ACCGATT CTTTGATC-T TTAGGA----
8I3    -GCAAA-TAT TAATTTTAAC CC-ACCGATT CTTTGATC-T TTAGGA----
6I3    -GCAAA-TAT TAATTTTAAC CC-ACCGATT CTTTGATC-T TTAGGA----
16I3   -GCAAA-TAT TAATTTTAAC CC-ACCGATT CTTTGATC-T TTAGGA----
1I2    -GTAAA-TAT TAATTTTAAC CC-ACCGATT CTTTGATC-T TTAGGA----
17I2   -GTAAA-TAT TAATTTTAAC CC-ACCGATT CTTTGATC-T TTAGGA----
5I2    -GTAAA-TAT TAATTTTAAC CC-ACCGATT CCTTGATC-T TTAAGA----
6I2    -GTAAA-TAT TAATTTTAAC CC-ACCGATT CTTTGATC-T TTAGGA----
18I    -GTAAA-TAT TAATTTTAAC CC-ACCGATT CTTTGATC-T TTAGGA----
13I    -GTAAA-TAT TATTTTTAAC CC-ACCGATT CTTTGATC-T TTAGGA----
20I    -GTAAA-TAT TAATTTTAAC CC-ACCGATT CTTTGATC-T TTAGGA----
12I    -GTATA-TAT TAATTTTAAC CC-ACCGATT CTTTGATC-T TTAGGA----
7I2    -GTAAA-TAT TAATTTTAAC CC-ACCGATT CTTTGATC-T TTAGGA----
7F2    -GCAAA-TAT TAATTTTAAC CC-ACCGATT CTTTGATC-T TTAGGA----
5F2    -GCAAA-TAT TAATTTTAAC CC-ACCGATT CTTTGATC-T TTAGGA----
12F2   -GCAAA-TAT TAATTTTAAC CC-ACCGATC CTTAGATC-T TTAGGA----
6F2    -GCAAA-TAT TAATTTTAAC CC-ACCGATT CTTTGATC-T TTAGGA----
3F2    -GCAAA-TAT TAATTTTAAC CC-ACCGATT CTTTGATC-T TTAGGA----
1F2    -GCAAA-TAT TAATTTTAAC CC-ACCGATT CTTTGATC-T TTAGGA----
1UK    -ACAAA-TAT TAATTTTAGT CC-ACCTATT TTATGATT-T TTAGGA----
8UK    -ACAAA-TAT TAATTTTAGT CC-ACCTATT TTATGATT-T TTATGA----
9UK    -ACAAA-TAT TAATTTTAAT CC-ACCTATT TTGTGATT-G TTAGGA----
10UK   -ACAAA-TAT TAATTTTAGT CC-ACCTATT TTATGATT-G TTAGGA----

9I2    TTTATTTTTT TTAA---TTT ACAATTGGTG G-AATTA-CA GGGATTA--T
11I2   TTTATTTTTT TTAA---TTT ACAATTGGTG G-AATTA-CA GGGATTA--T
10F1   TTTATTGTTT TAAA---TTA ACAATTGGTG G-AACTA-AA GGGATTT--T
2F1    TTAATTTTTT TTAA---TTT ACAAATTGTG G-AATTA-CA GGGATAA--T
15F1   TTTATTTCTT TAAA---TTT ACAATAGATG G-AACTA-CA GGGGTTC--T
18F1   TTTATTTTTT TTAA---TTT ACAATTGGTG G-AATTA-CA GGGATTA--T
5I3    TTTATTTTTT TTAA---TTT ACAATTGGTG G-AATTA-CA GGGATTA--T
2I3    TTTATTTTTT TTAA---TTT ACCATTGGTG G-AATTA-AA GGGATTT--T
1I3    TTTATTTTTT TTAA---TTT ACAATTGGTG G-AATTA-CA GGGATTA--T
10I3   TTTATTTTTT TTAA---TTT ACAATTGGTG G-AATTA-CA GGGATTA--T
7I3    TTAATTTTTT TTAA---TTT ACAATTGGTG A-AATTA-CA GGGATTA--T
12I3   TTAATTTTTT TTAA---TTT ACAATTGGTG G-AAATA-AA GGGATTA--T
8I3    TTTATTTTTT TTAA---TTT ACAATTGTTG G-AATAA-CA GGGATTA--T
6I3    ATTATTTTTT TTAA---TTT ACAATTGGTG G-AATAA-CA GGGATTA--T
16I3   TTTATTTTTT TTAA---TTT ACAATTGGTG G-AATTA-CA GGGATTA--T
1I2    ATTATTTATT TTAA---TTT ACGATAGGTG G-AATTT-CA GGGATTT--T
17I2   ATTATTTTTT TTAA---TTT ACAATTGGTG G-AATAA-CA GGGATTT--T
5I2    ATTATTTTTT TTAA---TTT ACAATAGGTG G-AATAA-CA GGGATTA--T
6I2    TTTATTTTTT TTAA---TTT ACAATAGGTG G-AATAA-CA GGGATTA--T
18I    TTTATTTTTT TTAA---TTT ACAATTGGGG G-AATTA-CA GGGATTA--T
13I    TTTATTTTTT TTAA---TTT ACAATTGGGG G-AATTA-CA GGGATTA--T
20I    TTTATTTTTT TTAA---TTT ACAATTGGGG G-AATTA-CA GGGATTA--T
12I    TTTATTTTTT TTAA---TTT ACAATTGGTG G-AATTA-CA GGGATTA--T
7I2    TTTATTTTTT TTAA---TTT ACAATTGGTG G-AATTA-CA GGGATTA--T
7F2    TTTATTTTTT TTAA---TTT ACAATTGGTG G-AATTA-CA GGGATTT--T
5F2    TTTATTTTTT TTAA---TTT ACAATTGGTG G-AATTA-CA GGGATTA--T
12F2   TTTATATTTT TTAA---TTT ACAATGGGTG G-AATTA-CA GGGATTA--T
```

Fig. 1 continued

```
  6F2   TTTATTTTTT TTAA---TTT ACAATTGGTG G-AATTA-CA GGGATTA--T
  3F2   TTTATTTTTT TTAA---TTT ACAATTGGTG G-AATTA-CA GGGATTA--T
  1F2   TTTATTTTTT TTAA---TTT ACAATTGGTG G-AATTA-CA GGGATTT--T
  1UK   TTTATTTTGT TAAT---TTT ACTATCGGCG G-AATTA-CA GGAATGA--T
  8UK   TTTATTTTGT TAAT---TTT ACTATCGGCG G-AATTA-CA GGAATGA--T
  9UK   TTTAGTCCGT AAAA---TCC ACTGTGTGGA A-TCTCT-CT GAGAGGA--T
 10UK   TTTATTCCGT AAAT---TTA ACTATAGGCG G-AATTA-CC GGAATGA--T

  9I2   TTTAG-CCAA T-TTC--AAG AA--TTGATA T-TATTTTTA CA-TGATA-C
 11I2   TTTAG-CCAA A-TTC--AAG AA--TTGATA T-TATTTTTA CA-TGATA-C
 10F1   TTTAC-CCAA A-TTC--AAG AA--TTGGAA T-TATGTTAA TA-TGATA-C
  2F1   TTTAT-CCAA A-TTC--AAG GA--TTGGAA T-TATATTAA TG-TGAAA-C
 15F1   TTTAC-CCAA A-TTC--AAG AA--TTAGAA T-TGTTTTAA CA-TGATA-C
 18F1   TTTAG-CCAA A-TTC--AAG AA--TTGATA T-TATTTTAA CA-TGATA-C
  5I3   TTTAG-CCCA A-TTC--AAG AA--TTGATA T-TATTTTAA CA-TGATA-C
  2I3   TTTAG-CCCA T-TTC--AAG AA--TTGATA T-TATTTTAA CA-TGATA-C
  1I3   TTTAG-CCCA A-TTC--AAG AA--TTGATA T-TATTTTAA CA-TGATA-C
 10I3   TTTAG-CCCA A-TTC--AAG AA--TTGATA T-TATTTTAA CA-TGATA-C
  7I3   TTTAG-CCCA A-TTC--AAG AA--TTGATA A-TATTTTAA CT-TGATA-C
 12I3   TTTAG-CCCA A-TTC--AAG AA--TTGATA A-TATTTTAA CT-TGAAA-C
  8I3   TTTAG-CCCA A-TTC--AAG AA--TTGATA A-TATTTTAA CT-TGATA-C
  6I3   TTTAG-CCAA A-TTC--AAG AA--TTGATA T-TATTTTAA CA-TGATA-C
 16I3   TTTAG-CCAA A-TTC--AAG AA--TTGATA T-TATTTTAA CA-TGATA-C
  1I2   TTTAG-CCGA A-TTC--AAG AA--TTGAAA T-TATTTTTA CA-TGGTA-C
 17I2   TTTAA-CCGA A-TTC--AAG AA--TTGATA T-TATTTTTA CA-TGGTA-C
  5I2   TTTAC-CCAA A-TTC--CAG AA--TTGAAA T-TATTTTTA CA-TGAAC-C
  6I2   TTTAG-CCAA A-TTC--AAG AA--TTGATA T-TATGTTTA CA-TGATA-C
  18I   TTTAG-CCAA A-TTC--AAG AA--TTGATA T-TAATTTTA CA-TGAAA-C
  13I   TTTAG-CCAA A-TTC--AAG AG--TGAATA T-TATTTTTA CA-GGAAA-C
  20I   TTTAG-CCAA A-TTC--AAG AA--TTGATA T-TATTTTTA CA-GGAAA-C
  12I   TTTAG-CCAA A-TTC--AAG AA--TTGATA T-TATTTTTA CA-TGAAA-C
  7I2   TTTAG-CCAA A-TTC--AAG AA--TTGATA T-TATTTTTA CA-TGATA-C
  7F2   TTTAG-CCAA A-TTC--AAG AA--TTGATA T-TATTTTTA CA-TGATA-C
  5F2   TTTAG-CCAA T-TCC--AAG AA--TTGATA T-TATTTTTA CA-GTCGA-C
 12F2   TTTAG-CCAA A-TTC--AAG AA--TGAATA T-TATATTTA CA-TGTAA-T
  6F2   TTTAG-CCAA A-TTC--TAG AA--TGAATA T-TATATTTA CA-GGAGA-C
  3F2   TTTAG-CCAA A-TTC--AAG AA--TTGATA T-TATTTTTA CA-TGATA-C
  1F2   TTTAG-CCAA A-TTC--AAG AA--TTGATA T-TATTTTTA CA-TGATA-C
  1UK   TTTAG-TCAA T-TCC--AGG AA--TGATCA T-TATTCTTC CA-TGATA-C
  8UK   TTTAG-TTCA A-TCC--AGG AA--TGATCT T-ATTCTTCC CA-TGATA-C
  9UK   TTTGA-TCAA C-TCC--TGA CA--TAGTGA G-CACTCCTC GA-GAATA-C
 10UK   TTTGA-TCAG C-TCC--AGA AC--TGGTCA T-TATTCCTC AA-TGATA-C

  9I2   TTA-CTATG- TA
 11I2   TTA-CTATG- TA
 10F1   CCA-ACAGG- TA
  2F1   CTA-CTATG- TA
 15F1   TTA-CCATG- TA
 18F1   TTA-CTATG- TA
  5I3   TTA-CTATG- GA
  2I3   TTA-CTATG- GA
  1I3   TTA-CTATG- GA
 10I3   TTA-CTAGG- GA
  7I3   TTA-CTATG- GA
 12I3   TTA-CTATG- GA
  8I3   TAA-CTATG- GA
  6I3   TTA-CTATG- GA
 16I3   TAA-CCATG- GA
  1I2   CTA-ATATG- GA
 17I2   CAA-ACCTG- GA
  5I2   TAA-CCAGG- GA
  6I2   TTA-CTATG- GA
  18I   TTA-CAAGG- GA
  13I   TTA-CAAGG- GA
  20I   TTA-CAAGG- GA
  12I   TTA-CAAGG- GA
  7I2   TTA-CTATG- TA
  7F2   TTA-CTAGG- CT
  5F2   AAT-GTAGG- CT
 12F2   TTA-CTCGG- CT
  6F2   TTA-CTATG- CT
  3F2   CTA-CTATG- CT
  1F2   TTA-CTATG- TA
  1UK   TTA-TTATG- TG
  8UK   TTA-TTATG- TG
  9UK   TTA-TTATG- TA
 10UK   TTA-TTAAG- TT
```

shows 55.3% of transitions (8 C/T, 3 T/C, 6 A/G, and 4 G/A) and 44.76% of transversions (4 T/A, 2 A/T, 1 A/C, 2 C/A, 5 G/T, 1 T/G, and 2 G/C) with respect to the other three samples.

The alignment of UK with F2 samples shows 34.8% of nucleotides variations between them and in 48% of these variations the UK samples have unique substitutions when compared to the F2 samples (52.5% of transitions and 65.6% of transversions).

The alignment of the UK samples with the F1 samples reveals 38.3% of nucleotide variations between them with 25% of transitions and 36.4% of transversions as unique substitution. Interestingly as shown on the tree the F1 samples from France are closer to the UK samples compared to the other French samples (F2). This was not expected as F1 and F2 samples are coming from two French farms from the same region on the West coast but still near the border with UK.

The alignment of UK samples with I samples shows a 34.8% of nucleotides variations, while with I2 samples there were a 34% of nucleotides variations. Finally with the I3 samples the percentage of nucleotides variations was 32.8%.

In Table 1 we report the number of transitions and transversions of UK samples compared to Italian and French samples.

French samples

Within the F2 samples the percentage of nucleotides variations was 8.5% and in 38.7% of these variations all the samples have the same nucleotides while one sample (12F2) had five transitions (2 T/C, 2 C/T, 1 A/G) and ten transversions (2 C/A, 2 G/T, 1 T/G, 2 T/A, 2 A/T, and 1 G/T). Within the F1 samples there were 14.2% of nucleotides variation and for one samples (10F1) there were three transitions (1 G/A and 2 C/T) and thirteen transversions (2 T/A, 3 C/A, 1 C/G, 3 G/T, 2 A/T and 2 A/C).

The alignment of F1 and F2 samples shows a 21.4% of variability with five A/T, 1 C/A, 1 T/G, 1 G/T, 1 G/A, 1 C/T, and 1 T/C significant nucleotides substitutions between them.

The DNA alignment shows 15.3, 12.3 and 13.1% variations between the F2 samples and the I, I3 and I2 farms, respectively. It also shows 21.6, 17.5 and 19.7% variations between the F1 farm samples and the I, I3, and I2 farms, respectively.

Table 1 Number of transitions and transversions between the UK farm and the Italian and French farms

UK farm variation	Number
T/C	12
C/T	16
G/A	9
A/G	10
G/T	9
T/G	10
T/A	15
A/T	13
A/C	2
C/A	10
C/G	6
G/C	1

Italian samples

The percentage of variation was 6.3% for the I samples, 4.9% for I3 samples and 7.7% for I2 samples. The three Italian farms show a 17.3% of variation between them. The I samples show eight important substitution compared to the other Italian samples (2A/C, T/A, G/T, T/G, C/A, and 2A/T). The I3 samples show one transition (C/T) and one transversion (A/T) compared to I and I2 samples.

Amino acid sequences

The sequences could be translated in amino acid sequences without any stop codons. The translated sequences gave 116–119 amino acid, on which 87 (75%) were conserved residues for I2 samples. The conserved residues in the F1 samples were 83 (71.5%), while for the F2 were 102 (85.7%). For UK samples the conserved residues were 82 (70.7%), for I samples were 99 (84.6%) and for I3 samples the conserved residues were 102 (87.9%).

Phylogenetic analysis

These results show that the mitochondrial COI gene of *D. gallinae* is suitable for phylogenetic analysis at the intra-specific level (Lee et al. 1999). Thus, NJ tree was constructed using the parameter values estimated above as reported in fig. 2.

The NJ tree shows a similar phylogeny among all UK samples and a more distant relationship with the Italian and French samples. All the UK samples are grouped in a single clade, which is supported by the bootstrap value (100%) showing that the strains used in this study belonged to a single lineage.

It shows that substitutions were greater in the North of Europe with UK having the highest rate of substitutions then F1 in France while samples in Italy showed little variation between them.

Discussion and conclusion

In this work, we studied the phylogeny of *D. gallinae* populations come from European countries, using the mitochondrial COI gene sequences. The mitochondrial DNA genes, especially the COI gene has been used extensively in phylogenetic studies due to the ease of primer design and its range of phylogenetic signal. The rate of evolution in this gene is also sufficiently rapid to allow the discrimination at the species level and the identification of cryptic species (Hebert et al. 2004), and has been used in establishing host plant associated genetic differentiation (Brunner et al. 2004).

The four UK sequences are showing greater diversity between themselves than the nine samples from the Italian I3 farm for instance. It highlights the level of environmental pressure in each country. It is also interesting to observe a North–South variation between the nucleotide changes and the polymorphism between the two French farms situated in the same region. Farmers stated that OPs, amidine and pyrethroid-based acaricides were the most commonly used; however, specific data on acaricide products used on each farm are not available which could impact on acaricide resistance and chemical pressure on the mite populations. Considering that all the samples are coming from cage systems it is possible that even more variability could exist between mite populations present in barn, free range and organic systems due to different control methods used in such alternative poultry

Fig. 2 Phylogenetic tree of 34 strains of *Dermanyssus gallinae*. The tree was constructed as described in "Materials and methods" using the neighbour joining boostrap support when Kimura's two parameter distance is used

systems, greater possibilities for mites to hide and avoid chemical control or poultry contact with wild birds bringing their own *D. gallinae* populations. Phylogenetic studies could help to address failure in acaricide control methods if mite populations have diverged.

Acknowledgments This work was partially supported by a research grant (Ricerca Corrente 2006–Istituto Zooprofilattico Sperimentale Della Puglia e Basilicata) from the Ministry of Health, Italy, by a grant (British–Italian Partnership Program) from the British Council- CRUI and by the European Commission through the STREP project "RESCAPE", contract no. 036018, under the 6th Framework Programme, priority 5, food quality and safety.

References

Ballard JWO, Rand DM (2005) The population biology of mitochondrial DNA and its phylogenetic implications. Annu Rev Ecol Evol Syst 36:621–642. doi:10.1146/annurev.ecolsys.36.091704.175513

Brannstrom S, Morrison DA, Mattsson JG, Chirico J (2008) Genetic differences in internal transcribed spacer 1 between *Dermanyssus gallinae* from wild birds and domestic chickens. Med Vet Entomol 22:152–155. doi:10.1111/j.1365-2915.2008.00722.x

Brockis DC (1980) Mite infestations. Vet Rec 107:315–316

Brunner PC, Chatzivassilio EK, Katis NI, Frey JE (2004) Host associated genetic differentiation in *Thrips tabaci* (Insecta: Thysanoptera), as determined from mtDNA sequence data. Heredity 93:364–370. doi: 10.1038/sj.hdy.6800512

Chauve C (1998) The poultry red mite *Dermanyssus gallinae* (De Geer, 1778): current situation and future prospects for control. Vet Parasitol 79:239–245. doi:10.1016/S0304-4017(98)00167-8

Chirico J, Eriksson H, Fossum O, Jansson D (2003) The poultry red mite, *Dermanyssus gallinae*, a potential vector of *Erysipelothrix rhusiopathiae* causing erysipelas in hens. Med Vet Entomol 17:232–234. doi: 10.1046/j.1365-2915.2003.00428.x

Cruickshank RH (2002) Molecular markers for the phylogenetics of mites and ticks. Syst Appl Acarol 7: 3–14

Desloire S, Valiente Moro C, Chauve C, Zenner L (2006) Comparison of four methods of extracting DNA from *Dermanyssus gallinae* (Acari: Dermanyssidae). Vet Res 37:725–732. doi:10.1051/vetres:2006031

Hebert PDN, Penton EH, Burns JM, Janzen DH, Hallwachs W (2004) Ten species in one: DNA barcoding reveals cryptic species in the neotropical butterfly *Astraptes fulgerator*. Proc Natl Acad Sci USA 101:14813–14817. doi:10.1073/pnas.0406166101

Hills DM, Dixon MT (1991) Ribosomal DNA: molecular evolution and phylogenetic inference. Q Rev Biol 66:411–453. doi:10.1086/417338

Hinomoto N, Takafuji A (2001) Genetic diversity and phylogeny of the Kanzawa spider mite, *Tetranychus kanzawai*, in Japan. Exp Appl Acarol 25:355–370. doi:10.1023/A:1017934218898

Hinomoto N, Osakabe M, Gotoh T, Takafuji A (2001) Phylogenetic analysis of green and red forms of the two-spotted spider mite, *Tetranychus urticae* Koch (Acari: Tetranychidae), in Japan, based on mitochondrial cytochrome oxidase subunit I sequences. Appl Entomol Zool (Jpn) 36:459–464. doi: 10.1303/aez.2001.459

Kambhampati S, Rai K (1991) Variation in mitochondrial DNA of *Aedes* species (Diptera: Culicidae). Evol Int J Org Evol 45:120–129. doi:10.2307/2409487

Kirkwood A (1963) Longevity of the mites *Dermanyssus gallinae* and *Liponyssus sylviarum*. Exp Parasitol 14:358–366. doi:10.1016/0014-4894(63)90043-2

Lee ML, Suh SJ, Kwon YJ (1999) Phylogeny and diagnostic markers of six Tetranychus species (Acarina: Tetranychidae) in Korea based on the mitochondrial cytochrome oxidase subunit I. J Asia-Pacific Entomol 2:85–92. doi:10.1016/S1226-8615(08)60035-7

Mangold AJ, Barques MD, Mas-Coma S (1998) Mithocondrial 16S rDNA sequences and phylogenetic relantionships of species of Riphicephalus and other tick genera among Metastriata (Acari: Ixodidae). Parasitol Res 84(6):478–484. doi:10.1007/s004360050433

Marangi M, Cafiero MA, Capelli G, Camarda A, Sparagano OAE, Giangaspero A (2008) Evaluation of the poultry red mite (*Dermanyssus gallinae*, *Acarina*: *Dermanyssidae*) susceptibility to some acaricides in a field population from Italy. Exp Appl Acarol 17. doi:10.1007/s10493-008-9224-0

Navajas M, Gutierrez J, Bonato O, Bolland HR, Mapangoudivassa S (1994) Intraspecific diversity of the cassava green mite *Mononycellus progresivus* (Acari, Tetranychidae) using comparisons of mitochondrial and nuclear ribosomal DNA-sequences and cross-breeding. Exp Appl Acarol 18:351–360. doi:10.1007/BF00116316

Puterka GJ, Black WCIV, Steiner WM, Burton RL (1993) Genetic variation and phylogenetic relationships among worldwide collections of the Russian wheat aphid, *Diuraphis noxia* (Mordvilko), inferred from allozyme and RAPD-PCR markers. Heredity 70:604–618. doi:10.1038/hdy.1993.87

Ros VID, Breeuwer JAJ (2007) Spider mite (Acari: Tetranychidae) mitochondrial COI phylogeny reviewed: host plant relationships, phylogeography, reproductive parasites and barcoding. Exp Appl Acarol 42:239–262. doi:10.1007/s10493-007-9092-z

Simon C, Frati F, Beckenbach A, Crespi B, Liu H, Flook P (1994) Evolution, weighting, and phylogenetic utility of mitochondrial gene sequences and a compilation of conserved polymerase chain reaction primers. Ann Entomol Soc Am 87:651–701

Sparagano O, Pavlicevic A, Murano T, Camarda A, Sahibi H, Kilpinen O, Mul M, van Emous R, le Bouquin S, Hoele K, Cafiero MA (2008) Prevalence and key figures for the poultry red mite *Dermanyssus gallinae* infections in poultry farm systems. Exp Appl Acarol. doi:10.1007/s10493-008-9233-z

Sperling FAH, Spence JR, Andersen NM (1997) Mitochondrial DNA, allozyme, morphology, and hybrid compatibility in *Limnoporus* water striders (Heteroptera: Gerridae): do they all track species phylogenies? Ann Entomol Soc Am 90:401–415

Sullivan J, Abdo Z, Joyce P, Swofford DL (2005) Evaluating the performance of a successive-approximations approach to parameter optimization in maximum-likelihood phylogeny estimation. Mol Biol Evol 22:1386–1392. doi:10.1093/molbev/msi129

Swofford DL (2002) PAUP*, Phylogenetic analysis using parsimony (*and other methods). Sinauer Associates, Sunderland

Swofford DL, Sullivan J (2003) Phylogeny inference based on parsimony and other methods using PAUP*. In: Salemi M, Vandamme A-M (eds) The phylogenetic handbook. A practical approach to DNA and protein phylogeny. Cambridge University Press, Cambridge

Thompson JD, Gibson TJ, Plewniak F, Jeanmougin F, Higgins DG (1997) The CLUSTAL_X windows interface: flexible strategies for multiple sequence alignment aided by quality analysis tools. Nucleic Acids Res 25:4876–4882. doi:10.1093/nar/25.24.4876

Toda S, Osakabe M, Komazaki S (2000) Interspecific diversity of mitochondrial COI sequences in Japanese *Panonychus* species (Acari: Tetranychidae). Exp Appl Acarol 24:821–829. doi:10.1023/A:1006484018547

Valiente Moro C, Chauve C, Zenner L (2007) Experimental infection of *Salmonella enteritidis* by the poultry red mite, *Dermanyssus gallinae*. Vet Parasitol 146:329–336. doi:10.1016/j.vetpar.2007.02.024

Valiente Moro C, De Luna C, Guy JH, Sparagano OAE, Zenner L (2008) The poultry red mite, *D. gallinae*, a potential vector of pathogenic agents. Exp Appl Acarol. doi:10.1007/s10493-009-9248-0

Zeman P, Stika V, Skalka B, Bártík M, Dusbábek F, Lávicková M (1982) Potential role of *Dermanyssus gallinae* De Geer, 1778 in the circulation of the agent of pullurosis-typhus in hens. Folia Parasitol (Praha) 29:371–374

Monitoring of *Dermanyssus gallinae* in free-range poultry farms

L. Zenner · G. Bon · C. Chauve · C. Nemoz · S. Lubac

Originally published in the journal Experimental and Applied Acarology, Volume 48, Nos 1–2, 157–166.
DOI: 10.1007/s10493-009-9253-3

Abstract Various methods for monitoring *Dermanyssus gallinae* infestations within free-range egg production units were compared. The study was carried out in five egg-producing free-range poultry buildings infested with *D. gallinae*. Each farm was divided into six zones (each zone including nest boxes, perches and duckboard) for placing two types of traps (corrugated cardboard and thick card traps) or examining dried droppings for presence of mites. Traps were removed 24 h later, placed into bags and mites were counted at the laboratory using binocular magnification. Droppings were also inspected by eye and mite numbers were estimated. All the methods used allowed us to detect mites although their efficacy differed. The number of mites collected was independent of the type of trap used. Examination of the droppings did not differentiate between buildings with differing mite populations. Placing traps in the nest boxes is a less reliable indicator than placing them on the perches. It appears that the most coherent method for evaluating the *D. gallinae* population within a free-range flock is to place thick card traps throughout the building, on perches favoured by birds.

Keywords Parasite · Red poultry mite · Trapping method · Laying hen · Alternative farming methods

L. Zenner (✉) · G. Bon · C. Chauve
Laboratoire de Biométrie et Biologie Evolutive, Ecole Nationale Vétérinaire de Lyon & Université de Lyon - CNRS UMR 5558, Ecole Nationale Vétérinaire de Lyon, 1 avenue Bourgelat, 69280 Marcy l'Etoile, France
e-mail: l.zenner@vet-lyon.fr

G. Bon · S. Lubac
Institut Technique de l'Aviculture Sud-Est, 23 rue Baldassini, 69007 Lyon, France

C. Nemoz
Hospices Civils de Lyon, Service de Biostatistique, Lyon 69424, France

C. Nemoz
Université de Lyon, Université Lyon I, Villeurbanne 69622, France

C. Nemoz
CNRS UMR 5558, Laboratoire Biostatistique Santé, Pierre-Bénite 69495, France

Introduction

There are two types of poultry farm raising laying hens. The battery farms include very standardised farms with several tens of thousands of battery-caged birds. The free-range farms include those farms where the laying hens are not caged, but are housed in deep litter or perchery/aviary systems with outside scratching area. These farms may include several thousand birds sharing the same space (Guillou 1988).

Dermanyssus gallinae (De Geer 1778), also known as the "red poultry mite", is the external parasite with the greatest impact on laying hens in Europe (Chauve 1998). This hematophagous acarian only parasites its host during blood meals. The rest of the time it lives hidden in the farm buildings. It is responsible for huge financial losses in French poultry farms breeding laying hens.

In battery farms, the economic impact of *D. gallinae* is partly due to the downgrading of eggs stained by the blood of squashed mites. This downgrading of egg quality has a financial impact with a proportion of second-quality eggs increasing by 2–14% (Van Emous et al. 2006). During the course of a massive infestation, the hens also suffer from anaemia which is further responsible for a reduction in their egg production and can lead to the death of the bird. Indeed, report of field outbreaks documenting severe anaemia and mortality invariably also include drop in egg production, regularly of 10–20% (Kirkwood 1967; Jungmann et al. 1970; Ambrosi and Flores 1972; Cosoroabă 2001; Cencek 2003).

In free-range poultry farms, the stress caused by this ectoparasite leads to pecking behaviour, often considered to have irreversible consequences and characterised by an increase in the number of pecks given to other birds (Roy et al. 2006); this can cause fatal lesions and increase the likelihood of cannibalism within the flock. It further causes an increase in the number of downgraded eggs when aggressive behaviour from the other birds causes the cloaca to become soiled with blood.

One study has found that 94% of battery flocks and 75% of free-range farms in south–east France had been treated with acaricides in an attempt to combat the proliferation of this acarian (Lubac et al. 2003). These treatments are expensive and further reduce the profit margins of the farmers.

One of the characteristics of infestation by red mites is that once they are present in a flock they are almost impossible to eradicate (Chauve 1998). However, while a low level infestation may be deemed acceptable since it causes only minimal losses, the difficulty for the farmer is to know exactly when is the best time to treat the flock as the adult mites are less than a millimetre in size and difficult to see with the naked eye. Furthermore, they hide away in cracks and crannies of the building, laying their eggs during the day, far away from the light (Collins and Cawthorne 1976).

Studies on the dynamics of *Dermanyssus* populations have already been carried out in battery farms (Levot 1991) and in perchery/aviary production units (Nordenfors and Chirico 2001; Chirico and Tauson 2002; Lundh et al. 2005), but never in free-range production units commonly found in France. However, it is important to assess, even semi-quantitatively, the mite population in such farms for at least two reasons: evaluation of control measures together with the optimal period for treatment and in situ studies of the parasite. The aim of our study was to compare the various methods available for quantifying parasitic infestation and to define those areas within a free-range production unit building best suited to testing.

Materials and methods

The farms

This study was carried out in five farm buildings (A, B, C, D and E) in four free-range farms breeding laying hens in the departments of the Ain, the Ardèche and the Rhône (France). Three of the farms raised organic chickens, and the fourth certified birds, meeting very precise specifications as to their feed etc. The time the pullets spent on the farm varied from 7 to 33 weeks.

Description of the traps used

Two types of trap were used (Fig. 1a, b, c). The first type, based on the description by Nordenfors and Chirico (2001), was made up of a 3 mm thick sheet of corrugated cardboard, measuring 10 × 7 cm. The cardboard was cut in such a way that the corrugations were open on the longer side. The second type of trap, inspired by Levot (1991), consisted of a 20 × 7 cm piece of thick cardboard (210 g/cm^2) folded in two and held together by two staples to give a trap measuring 10 × 7 cm. Both types of trap were protected from being attacked by the hens with a 1 mm thick auto-adhesive rigid plastic sheet.

Fig. 1 **a** Open thick carboard trap with *Dermanyssus* after 24 h of freezing; **b** Pair of corrugated cardboard and thick cardboard traps in a nest box; **c** Fixation of a trap on a perch; **d** Dried dropping sample showing *Dermanyssus* on the underside

Examination of the bird droppings

Dried dropping samples around 5 cm long were taken from the duckboards (Fig. 1d), and the undersides examined. Based on the number of mites found in the dropping samples and on the underside of the duckboard scores were given as follows: 0: no mites, 1: 1–20 mites, 2: 20–200 mites, 3: more than 200 mites. The final mark given to the sample was obtained from the average of the scores from four dropping samples taken in any one area of approximately one meter square.

Experimental design

The first experiment was to compare different methods for quantitative evaluation of the presence of *D. gallinae* in a flock. Two different type of trap were tested in the farms with different infestation burdens (corrugated cardboard and thick cardboard traps as described previously). They were placed in pairs, one of each type, at a distance of 20 cms in nest or on perches (Fig. 1b). They were fixed in the nest boxes using double-sided sticky tape, and at the top of the perches using metal wire. They were taken down 24 h later and immediately placed in hermetically sealed bags for transport to the laboratory where they were frozen at −20°C for 24 h. The number of mites was individually counted with the help of a binocular magnifier after opening the traps. Seventy-six pairs of traps were tested. Counting *Dermanyssus* from traps were also compared to examination of dropping. In this experiment, 498 values from the five farms were analysed. Finally, data from 174 traps placed on the perches and 174 traps in the nest boxes were compared to evaluate the place to put the traps.

The second experiment compared the performance of the various assessment methods when there is a sudden fall in the population of *Dermanyssus*. To obtain this variation in infestation burdens, we have taken advantage of a trial of various acaricides in these farms. We placed traps in nest boxes and perches and examined droppings before (D0) and 7 days after (D7) the use of an acaricide and compared the differences in results from traps or dropping examination made exactly at the same place. Data from 30 traps in nest boxes, 30 traps on perches and 120 examination of dropping were used in this experiment.

The last experiment aimed to evaluate variation of infestation burdens observed in different areas of one building. We have defined a sampling zone which could be found in any free-range poultry farms regardless of its architecture as a area including nest boxes, perches and duckboards at proximity. For each farms, six different sampling zones were determined. For each sampling zone, two traps were placed (one in nest boxes and one on perches) and four examination droppings were made.

Statistical data analysis

Means comparisons were carried out using an ANOVA and the Statview programme in comparing: (1) the two type of traps, (2) the relative sensitivity of using traps or examining droppings to determine infestation levels, (3) the traps placed on the perches or in the next boxes, (4) the data obtained before and after use of an acaricide and (5) the data obtained from six areas of the same building. The correlation coefficient from data of the corrugated cardboards and thick cardboards was calculated using Spearman's

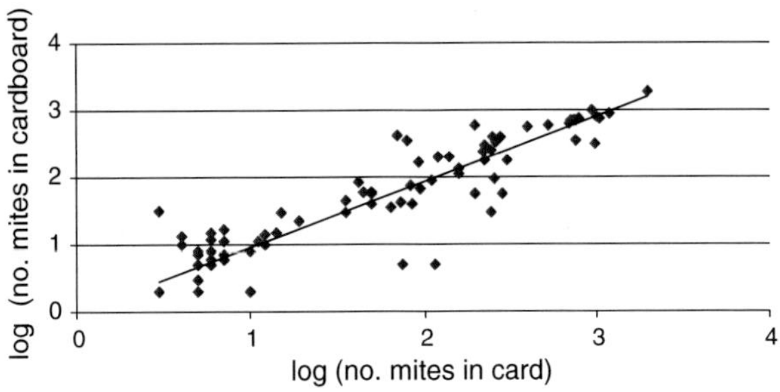

Fig. 2 Comparison of the number of red poultry mites present in the thick card and corrugated cardboard traps (Spearman's correlation $r^2 = 79.9\%$)

correlation coefficient. All information concerning the data tested was added to the experimental design section.

Results

Comparison of the different methods for quantitative evaluation of the presence of *Dermanyssus gallinae* in a flock

– Comparison of the two types of trap:

Data was obtained from 76 pairs of traps and the number of mites found in the corrugated cardboard and thick cardboard traps studied. The average number of mites collected was 239.5 ± 375 mites for the corrugated cardboard traps and 260.6 ± 471 mites for the cardboard traps, with no significant difference between the two types of trap.

Furthermore, an important correlation in mite numbers was found when comparing the results obtained from each pair of traps (Fig. 2).

Statistical analysis of the results was carried out using Spearman's correlation method (Rho = 79.7% with $P < 0.0001$). The two variables are therefore strongly correlated which implies that using either type of trap to measure the mite population gives similar results. For practical reasons, we continued the study using only the thick card traps as these were easier to use.

– Comparison of the relative sensitivity of using traps or examining droppings to determine infestation levels.

All three methods (traps on the perches, traps in the nest boxes, examination of droppings) were used in the five buildings, all of which were infested to varying degrees. Four hundred and ninety-eight readings were used for this comparison. Univariate variance analysis carried out on our results show that, contrary to the traps, examination of the droppings did not allow differentiation between the various buildings (data not shown). Therefore it follows that examination of droppings is a less sensitive test than using traps.

– Comparison between the traps placed on the perches or in the nest boxes.

This comparison was carried out on 174 values for each of the two trap sites. The average number of mites (±Standard Deviation) obtained from the traps placed in the nest boxes was 64.3 (±165) and 364.2 (±490) for the traps placed on the perches. Within any one farm, the number of mites found in the traps on the perches is from 1.7 to 77 times that of the nest boxes (Fig. 3).

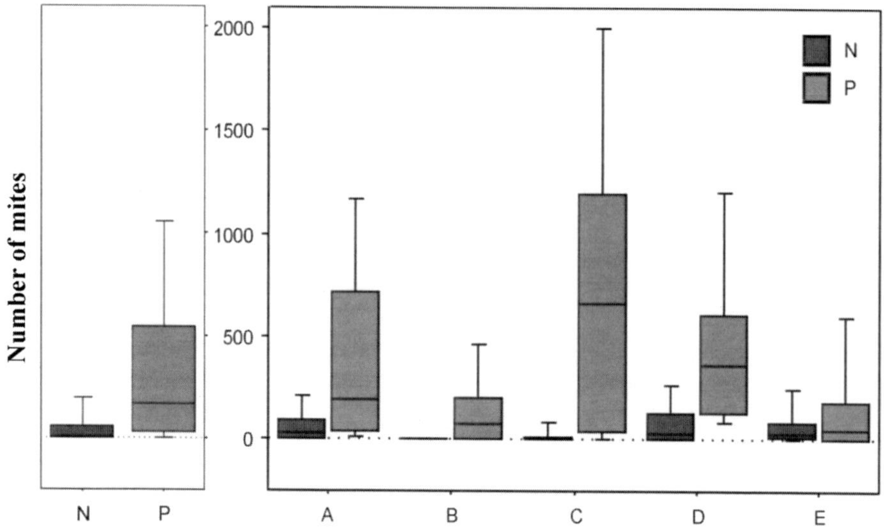

Fig. 3 Comparison of the number of red poultry mites collected from the nest boxes (N) and the perches (P). (1) overall and (2) by farm building (A, B, C, D, E) (shown as a boxplot-median, 25 and 75% percentiles and extremes)

Performance of the various assessment methods when there is a sudden fall in the population of *Dermanyssus gallinae*

A quantitative assessment of the number of mites was carried out using all three methods (traps on the perches, traps in the nest boxes, and examination of droppings) both before (D0) and 7 days after (D7) use of an acaricide in the production unit.

The data obtained from examination of the traps gave the greatest differences before and after treatment (Fig. 4). The number of mites found in the traps placed in the nest boxes showed a reduction of 82%, on the perches of 81% but only a reduction of 39% was noted from examination of the droppings. As *Dermanyssus* are mobile during the night, we can assume that treatment has the same impact on mites in the different places and this reinforces the reduced interest of using examination of droppings as an assessment method.

Comparison of the number of mites obtained from six areas of the same building

The lack of significance of the study of the droppings and the low number of mites collected in the nest boxes lead us to analysed only data from traps placed on perches. The number of mites collected from these traps was studied in relation to the six areas as defined in the Materials and methods section (Fig. 5).

Variations were noted in relation to the buildings studied: the results obtained from buildings B and E are similar in all zones, whereas in building C the difference ranges from 1 to 5. One area show there aren't any mites present in the trap (area 3, building C). But, univariate variance analysis carried out for each of these buildings did not allow us to show any significant difference between the six sampling areas.

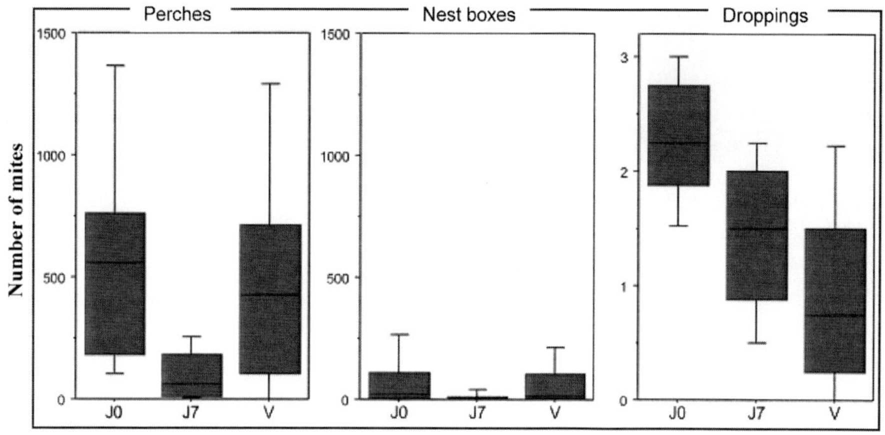

Fig. 4 Representation of average red poultry mite populations detected before (D0) and 1 week after (D7) application of an acaricide treatment. The variation (V) corresponds to the average of the observed differences for any given area between the populations before and after treatment (shown as a boxplot—median, 25 and 75% percentiles and extremes)

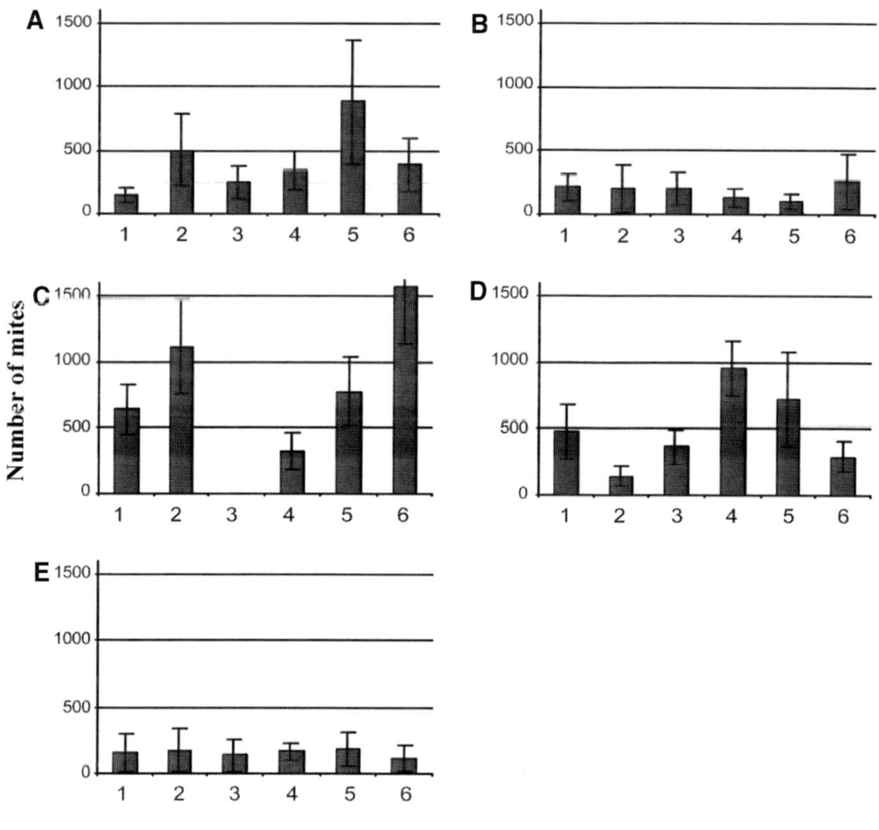

Fig. 5 Averages of red poultry mites collected from the perches in each of the six areas in the five buildings

Discussion

One of the problems faced in studying flocks with red mite infestation is evaluation of the parasite burden contained within the building and their distribution throughout the flock (Levot 1991; McGarry and Trees 1991; Nordenfords and Chirico 2001). From this point of view, it was therefore interesting to compare the various techniques available for quantitative evaluation of the mite population.

All the methods used: examination of droppings, traps in corrugated cardboard or thick cardboard, placed in nest boxes or on perches, allowed us to monitor the mites although their efficacy differed. Studying the two types of trap, whether in corrugated cardboard or in thick cardboard, showed that the number of mites collected was independent of the type of trap used. We preferred the thick card trap as it was easier to use. One of the inconveniences of these traps is their lack of resistance to humidity since simple contact with fresh droppings humidifies these traps and renderes them useless for mite collection. Any comparison with Nordenfors' study in Sweden (Nordenfors and Chirico 2001) is difficult even though he also used corrugated cardboard traps as the density of the hens per square metre in Swedish farms, which are of the caged type (14–20 birds per m^2) is well above that in the French farms we studied (6 birds per m^2). This could explain the average figure of 11,500 ± 1,590 mites obtained in Sweden. Furthermore, the traps were left in place for two nights which may also explain the higher number of mites reported by Nordenfors and Chirico. Examination of the droppings, a method routinely used by the farmers themselves and their technical experts, was shown to be unreliable as a method of assessing infestation as it did not allow us to differentiate between buildings with differing mite populations. The variations observed after a rapid fall in the mite population were also too small to consider examination of droppings as a reliable guide to the efficacy of an acaricide. However, this method remains a simple means of observing whether or not poultry red mites are present in the flock.

An interesting point is that of where to place the traps in the farm building. The quantitative difference in the number of mites collected from the traps placed on the perches or in the nest boxes shows that placing the traps in the nest boxes is a less sensitive indicator. One reason for this may be the shorter time that the birds spend in the nest boxes compared to on the perches. The presence of the parasite in any area of the farm building is strongly correlated with the behaviour of the birds themselves. We were further able to observe that the zero samples taken from the perches in building C, corresponded to the fact that the birds did not use these perches as they were some distance from the duckboard. It is therefore necessary to place the traps in areas actually used by the birds. Similarly, in buildings A, D and E, the proximity of the perches to the nest boxes is probably the explanation for the very large number of mites collected from the traps in these nest boxes.

One further question posed at the start of our study was whether the observation of several predetermined areas spread throughout the building, would allow us to evaluate the mite population in the entire building, regardless of its architecture. Each of the areas studied, defined by ourselves included elements common to all alternative poultry farms: the duckboard, the perches and the nest boxes. Unfortunately, the lack of significance from the study of the droppings and the low number of mites collected in the nest boxes rapidly narrowed our study area to finally include only the perches. The statistical study carried out on the data obtained from the perches allowed us to differentiate between the various study areas.

It therefore appears that the most coherent method for evaluating the *Dermanyssus* population within a flock is to place thick card traps throughout the building, on perches favoured by the birds. But as we reduced the sample size, using only those traps placed on the perches, it would be interesting to increase the number of traps, doubling it to 12 traps per building. In order to monitoring *D. gallinae* in poultry house premises, it would be important to optimise the number of traps to obtain reliable data. This Optimum Sample Size (OSS) had already been evaluated for litter beetles in poultry houses. Safrit and Axtell (1984) calculated this OSS at ten tube traps for *Alphitobius diaperinus* whereas Stafford et al. (1988) found OSS at 35 traps to monitor a high density of *Dermestes maculates*. In their study, Nordenfors and Chirico (2001) showed that the number of traps required to monitor *Dermanyssus* at any given level of variability decreased with higher mite densities.

This method gives a quantitative estimate of the mite population, although it in no way gives an actual figure for the number of *Dermanyssus* present in the flock. However, it does allow follow-up of the variations in the acarian population since we believe that the number of mites collected from the traps is certainly correlated to the *Dermanyssus* population present in the building. This method therefore allows for monitoring of the variations in the mite population during an acaricide treatment or during changes in environmental conditions within a building, such as differences in temperature or hygrometry. Incorporation of traps into pest control programmes is thus likely to result in a higher standard of management, involving quality assessment.

One possible use for this method would be in determining the threshold level for initiating control treatment. This threshold, to be defined, would depend both on the characteristics of the farm building, the evident distress levels of the hens and the efficacy of the product to be used in the treatment. The definition of such a threshold level would require long-term studies within a single farm building.

In conclusion, this study has clearly shown that it is possible to have an idea of the level of parasite infestation and to monitor it over time. This technique, even if it does not allow for determination of the exact number of red poultry mites present, does remain a valuable tool for understanding phenomena regarding variation of the mite population within a flock and permits monitoring of the efficacy of control treatments.

Acknowledgments We thank G. Martin for reading and correcting the manuscript.

References

Ambrosi M, Flores G (1972) La lotta contro *Dermanyssus gallinae* negli allevamenti avicoli. Nuova Vet 48:301–307

Cencek T (2003) Prevalence of *Dermanyssus gallinae* in poultry farms in Silesia region in Polland. Bull Vet Inst Pulawy 47:465–469

Chauve C (1998) The poultry red mite *Dermanyssus gallinae*: current situation and future prospects for control. Vet Parasitol 79:239–245. doi:10.1016/S0304-4017(98)00167-8

Chirico J, Tauson R (2002) Traps containing acaricides for the control of *Dermanyssus gallinae*. Vet Parasitol 110:109–116. doi:10.1016/S0304-4017(02)00310-2

Collins DS, Cawthorne RJG (1976) Mites in the poultry house. Agr Northen Ireland 51:358–366

Cosoroabă I (2001) Observation d'invasions massives par *Dermanyssus gallinae* (De Geer 1778), chez les poules élevées en batterie en Roumanie. Rev Méd Vét 152:89–96

Guillou M (1988) La poulette et la pondeuse d'œufs de consommation. In: Rosset R (ed) L'aviculture Française. Information technique des services vétérinaires, Paris, pp 297–318

Jungmann R, Ribbeck R, Eisenblätter S, Schematus H (1970) Zur Schadwirkung und Bekämpfung des *Dermanyssus gallinae*–und Federlingbefalls bei Legehennen. Monatsh Veterinarmed 25:28–32

Kirkwood A (1967) Anemia in poultry infested with red mite *Dermanyssus gallinae*. Vet Rec 80:514–516

Levot G (1991) Chemical control of *Dermanyssus gallinae* (Acarina: Dermanyssidae) on caged layer hens. Gen Appl Entomol 23:49–52

Lubac S, Dernburg A, Bon G, Chauve C, Zenner L (2003) Problématiques et pratiques d'élevage en poules pondeuses dans le sud est de la France contre les nuisibles: poux rouges et mouches. Comptes Rendus des cinquièmes Journées de la Recherche Avicole, Tours, pp 101–104

Lundh J, Wiktelius D, Chirico J (2005) Azadirachtin-impregnated traps for the control of *Dermanyssus gallinae*. Vet Parasitol 130:337–342. doi:10.1016/j.vetpar.2005.02.012

McGarry J, Trees A (1991) Trap perches to assess the activity of pyrethrins against the poultry red mite *Dermanyssus gallinae* in cage birds. Exp Appl Acarol 12:1–7. doi:10.1007/BF01204395

Nordenfors H, Chirico J (2001) Evaluation of a sampling trap for *Dermanyssus gallinae* (Acari: Dermanyssidae). J Econ Entomol 94:1617–1621

Roy L, Valiente-Moro C, Chauve C (2006) Pou rouge: diagnostic et lute contre l'infestation. Point Vet 266:48–52

Safrit RD, Axtell RC (1984) Evaluations of sampling methods for darkling beetles (*Alphitobius diaperinus*) in the litter of turkey and broiler houses. Poult Sci 63:2368–2375

Stafford KC, Collison CH, Burg JG, Cloud JA (1988) Distribution and monitoring lesser mealworms, hide beetles, and other fauna in high-rise, caged-layer poultry houses. J Agr Entomol 5:89–101

Van Emous RA, Fiks van Niekerk, Mull MF (2006) Red mites in theory and practice. *Praktijkrapport Pluimvee #17*. Animal Science Groupe, Lelystad, Netherland

Zenner L, Bon G, Dernburg A, Lubac S, Chauve C (2004) Preliminary studies of the monitoring of *Dermanyssus gallinae* in free-range poultry farms. Br Poult Sci 44:781–782

Preventing introduction and spread of *Dermanyssus gallinae* in poultry facilities using the HACCP method

Monique F. Mul · Constantianus J. M. Koenraadt

Originally published in the journal Experimental and Applied Acarology, Volume 48, Nos 1–2, 167–181.
DOI: 10.1007/s10493-009-9250-6

Abstract Preventing the establishment of ectoparasitic poultry red mite (*Dermanyssus gallinae*) populations is key in ensuring welfare and egg production of laying hens and absence of allergic reactions of workers in poultry facilities. Using the Hazard Analysis and Critical Control Point method, a panel of experts identified hazards and associated risks concerning the introduction and spread of this mite in poultry facilities. Together we provide an overview of possible corrective actions that can be taken to prevent population establishment. Additionally, a checklist of the most critical control points has been devised as management tool for poultry farmers. This list was evaluated by Dutch and British poultry farmers. They found the checklist feasible and useful.

Keywords Control · Corrective actions · Monitoring · Checklist · Risks · Farm processes

Introduction

Dermanyssus gallinae (De Geer 1778) is the most common ectoparasite in poultry. It belongs to the subclass Acari and is known under the common name poultry red mite (PRM) or chicken mite. Adult poultry red mites are on average 751 microns in length and 461 microns in width when engorged and are found in cracks and crevices within the poultry facilities in the vicinity of the hens. In these cracks and crevices the mites mate, deposit their eggs and molt. The life cycle of the mite contains five stages: egg, larva, protonymph, deutonymph, and adult (Wood 1917). The protonymph, deutonymph, and

M. F. Mul (✉)
Animal Production, Animal Sciences Group of Wageningen UR, P.O. Box 65, 8200 AB Lelystad, The Netherlands
e-mail: monique.mul@wur.nl

C. J. M. Koenraadt
Laboratory of Entomology, Department of Plant Sciences, Wageningen University, P.O. Box 8031, 6700 EH Wageningen, The Netherlands
e-mail: sander.koenraadt@wur.nl

adult feed on blood of poultry and other birds, but also of other animals, including humans (Sikes and Chamberlain 1954). The poultry red mite usually stays at the hen for a bloodmeal during the dark period. Within 8 weeks one female may produce an estimated amount of two and a half thousand offspring (Wood 1917).

Infestations with the poultry red mite can reduce the welfare of chickens, increase mortality and initiate allergic reactions of workers in the poultry facilities (Baselga et al. 1996; Chauve 1998; Kilpinen et al. 2005; Nordenfors 2000). The mite is a potential vector of various pathogens, such as *Salmonella* and the causative agents of fowl cholera, Newcastle disease and eastern equine encephalitis (Moro et al. 2005). Control of *D. gallinae* has become more difficult due to development of resistance to permethrin (Chauve 1998; Marangi et al. 2008; Nordenfors et al. 2001) and a ban on other acaricides in some countries such as carbaryl (a carbamate). Therefore, recent research has been focused on alternative control methods for *D. gallinae* such as the use of entomopathogenic fungi, silica, improved hygiene (e.g., heat cleansing, washing) and the use of natural predators (Gjevre, personal communication; Maurer and Hertzberg 2001; Maurer and Perler 2006; Nordenfors 2000; Steenberg et al. 2005). However, control could be more efficiently employed if prevention of the introduction and spread of *D. gallinae* was more rigorously enforced.

The Hazard Analysis and Critical Control Points (HACCP) system was introduced in the 1960s by NASA for design and manufacturing of food for spaceflights. Since then HACCP has been recognized internationally as a logical tool in the adaptation of traditional inspection methods to a modern, science-based, food safety system (Mayes 1993, 1998; Sun and Ockerman 2005). The advantages of such a structured and formalized concept was recognized and adapted for use in safeguarding animal welfare and health (Bonde and Sørensen 2004; Noordhuizen and Frankena 1999). For example, HACCP allows one to identify the risk factors for introduction and spread of bacteria and parasites such as *Salmonella* and *Toxoplasma gondii* (Kijlstra et al. 2004; Mul and Van der Gaag 2003). Moreover, it has been employed in risk assessment of the introduction of unwanted organisms on passenger ships (Mouchtouri et al. 2008) and it has been nominated for the prevention of fire ant introduction and spread in Australia (Rayment 2006). It is hypothesized that, in a similar way, the HACCP method can be used for risk factor assessment concerning introduction and spread of *D. gallinae* on poultry farms. Earlier work demonstrated that poultry red mites are considered the most severe and most frequent hazard in relation to 9 other poultry-related hazards such as pasteurellosis and cannibalism (Hegelund and Sørensen 2007). This paper further elaborates on the identification of the risk factors and critical control points, and presents suggestions for corrective actions. Based on this information and evaluation by farmers, we have developed a checklist that can be used to control *D. gallinae* more readily and effectively on poultry facilities.

HACCP development

HACCP is a method of controlling hazards and reducing risks. It comprises seven principles (Mayes 1998): (1) conduct a hazard analysis, (2) identify critical control points, (3) establish critical limits for each critical control point, (4) establish critical control point monitoring requirements, (5) establish corrective actions, (6) establish record keeping procedures, and (7) establish procedures to ensure that the HACCP system is working as intended (validation and verification). In order to apply this method to achieve insight in the hazards for introduction and spread of *D. gallinae* some adjustments were required. For example, instead of eradication or control of a critical control point (demand of HACCP), a reduction of the hazard had to be accepted and instead of analyzing the risk of a product we analyzed the risk of the potential (rate and extent) introduction or spread of *D. gallinae* in the poultry facility.

Hazard analysis

A hazard analysis is conducted by compiling a schedule of all farm processes. Due to the scarcity of quantitative information on epidemiological risk factors for introduction and spread of *D. gallinae* infestations, an expert opinion study was performed as others have in the past (Bonde and Sørensen 2004; Noordhuizen and Frankena 1999). Assessment of the risk of *D. gallinae* infestation and spread was judged by four experts. Three of them were experts in poultry husbandry and two of them were experts on *D. gallinae*. The farm processes under assessment were divided into 13 hazard categories (Table 1). In total 41 hazards were identified. Risk was calculated by multiplying likelihood and severity. Likelihood was classified into three categories: (1) occurring seldom or only theoretically, (2) occurring approximately once a year, and (3) occurring repeatedly or more than once a year throughout the year. Severity was classified as (1) low when only a single place in the poultry facility becomes infested with *D. gallinae*, (2) moderate, when the poultry facility becomes infested with *D. gallinae* at more than one location or (3) high, when *D. gallinae* infestation occurs at almost all places within the poultry facility. The panel of experts made several suggestions for possible corrective action. Details of the hazards are shown in Table 2.

Critical control point identification

A critical control point (CCP) is a step, point or procedure in any production procedure. This can also be applied to the egg production facility to identify and manage potential

Table 1 Overview of hazard categories, number of hazards and critical control points (CCPs) per category and associated total and average risk (details of the hazards are shown in Table 2)

Hazard category	Risk of introduction	Risk of spread	No. of hazards	No. of CCPs	Total risk score	Average risk score
Environment	x	x	7	5	28	4.0
Feed	x	x	4	1	10	2.5
Litter	x	–	1	1	3	3.0
Growing hens	x	–	4	3	23	5.8
Material/equipment	x	x	3	2	13	4.3
Manure	x	x	3	2	16	5.3
Eggs	x	x	4	4	21	5.3
Manure aeration	x	x	2	2	12	6.0
Cadavers	x	x	2	2	12	6.0
Visitors/external personnel	x	x	3	3	14	4.7
Poultry farmer/ employee	x	x	3	3	24	8.0
Ventilation	x	x	3	1	7	2.3
Unproductive hens	x	–	2	2	6	3.0

Table 2 Explanation of hazards and associated risks for introduction and spread of *Dermanyssus gallinae* in poultry systems

Hazard category	Hazard	Likelihood	Severity	Risk	Motivation	Corrective action
Environment	Introduction of mites via pests and vermin such as rats, mice, etc.	2	2	4	Poultry houses are attractive to rats, mice etc. because of the presence of food, water and opportunities for shelter. These animals are potential carriers of mites.	Professional pest/vermin control No storage in immediate vicinity poultry house Strip (2 m wide zone) along the poultry house free of vegetation Hard pavement (e.g., gravel or cobblestones) next to poultry house instead of grass
Environment	Introduction of mites into poultry house via wild birds	2	2	4	Mites may be present by a wild bird straying from the nest	Close poultry house for birds (bird-proof) Install springs on doors No shelter/runs for hobby poultry/birds next to poultry house
Environment	Introduction of mites via bird nests attached to the building	2	2	4	Many wild bird nests are infested with mites. After fledging of young birds in spring, mites starve and seek new hosts inside the poultry house.	Fill space underneath the (corrugated) roof to prevent breeding of birds No vegetation alongside poultry house that could house bird nests No nesting boxes against wall of poultry house
Environment	Introduction of mites into poultry house via pets	1	2	2	In the absence of birds or chickens, mites can chose dogs or cats as host, but the likelihood is considered low.	Do not allow pets in poultry house or associated spaces Install springs on outside doors
Environment	Spread of mites via rats, mice, flies, etc.	2	3	6	Pest animals such as rats and mice are potential carriers of mites. They can spread mites within a poultry house. Severity is higher, because mites have already been introduced before further spread occurs.	Professional pest/vermin control

Table 2 continued

Hazard category	Hazard	Likelihood	Severity	Risk	Motivation	Corrective action
Environment	Spread of mites via wild birds	2	3	6	Wild birds could carry mites and spread them within the poultry house. The likelihood depends on infestation pressure among wild birds. Wild birds could reach many areas within the poultry house.	If possible, remove wild birds from poultry house
Environment	Spread of mites via pets	1	2	2	Pets could carry and spread mites.	Do not allow pets in poultry house or associated spaces
Feed	Introduction of mites into poultry house via road transport	1	2	2	Feed trucks visit different farms and could theoretically carry mites from one farm to another. Likelihood and severity are low, because even if infested, it will be hard for mites to move from truck to poultry house	Keep distance between truck and poultry house
Feed	Introduction of mites into poultry house via feed bags	1	1	1	Although prepared clean from the factory, feed bags could be infested at visits to different farms during transportation.	Store feed bags in different area
Feed	Spread between barns via feed transport system	1	1	1	The feed transport system is a very low hazard for transport of mites between barns.	Seal holes where transport system enters/ leaves the barn with cement
Feed	Spread within barn via feeding system	2	3	6	Mites hide in (bent) edges/rims of the feeding system and may end up in the food or on the transporting chain and thus spread throughout the poultry house.	Treat feeding system with control agent Seal edges and rims of feeding system with silicone sealant
Litter	Introduction of mites in poultry house through scattering litter	1	3	3	Litter is clean and generally not a good hiding place for mites (low likelihood). However, if litter is infested, it will be spread throughout the entire poultry house (high severity)	Only use dry, clean litter

Table 2 continued

Hazard category	Hazard	Likelihood	Severity	Risk	Motivation	Corrective action
Growing hens	Introduction of mites with new flock of growing hens	3	3	9	Chickens from production farms are caught early in the morning. If the production farm is infested, some chickens could carry the mites to the laying facility.	More attention for prevention and monitoring of mites during growing phase Switch lights on 1 h before collection in battery systems Switch lights on 1 h before and switch to blue light during collection in free-range and aviary systems Mix garlic in drinking water several days before transport
Growing hens	Introduction of mites with containers/crates	3	3	9	A large proportion of mites that are on the chicken during collection, will move from the chicken and hide in containers/crates. These crates are placed throughout the entire poultry house when new flock is introduced (high severity).	Containers/crates should be cleaned
Growing hens	Introduction of mites by driver	2	1	2	Mites could move from chickens onto the truck driver.	Driver should not enter poultry house
Growing hens	Introduction of mites by employees	1	3	3	Teams that introduce new flocks visit different farms and could thus carry an infestation from one location to the other.	Personnel should be clean upon arrival at the farm (clean clothing)
Material/ equipment	Introduction of mites via material or equipment	1	1	1	New equipment or maintenance tools will not have mites.	Equipment should be cleaned
Material/ equipment	Spread of mites between barns via shared material/equipment	3	2	6	Material or equipment from another barn could spread mites. Maintenance of the poultry house occurs regularly and at various locations leading to a relatively high risk.	Tools that could harvest mites such as a broom should be kept in the same barn and regularly cleaned or disinfected

Table 2 continued

Hazard category	Hazard	Likelihood	Severity	Risk	Motivation	Corrective action
Material/ equipment	Spread of mites within a barn via material/equipment	3	2	6	Infested equipment could spread mites within a barn. Cleaning takes place regularly, leading to relatively high risk.	Regularly clean equipment that may harvest mites Keep equipment located centrally and attached to a board
Manure	Introduction of mites via manure conveyor belts	1	1	1	If manure storage is not close to the barn, the likelihood and severity are considered very small.	Clean manure conveyor belt regularly
Manure	Spread of mites between barns via cross-belts	3	2	6	Many mites can be encountered on manure conveyor belts. Between barns mites could spread where conveyor belts cross.	Regularly treat transfer points between manure belts and cross-belts with control agent
Manure	Spread of mites in poultry house via manure belts	3	3	9	In practice, many mites are encountered on manure belts. Mites could also hide on the inside and thus spread throughout the entire poultry house.	Automated control during transport of manure
Eggs	Introduction of mites via egg containers and pallets	3	1	3	Containers and pallets move from farm to farm. Containers/pallets have lots of hiding opportunities. However, these materials will not move further than the egg collection compartment on the farm (low severity).	Egg containers and pallets should be cleaned and disinfected by the packing station
Eggs	Introduction of mites via egg trays	3	1	3	Used trays have lots of hiding opportunities. However, trays will not move further than the egg collection compartment on the farm (low severity).	Use new trays Packing station should clean plastic egg trays
Eggs	Spread of mites between barns via egg cross-belts	3	2	6	Frequently, mites are encountered on eggs and egg remains. Cross-belts between main belts may carry mites from barn to barn. Especially transfer points between cross and main belts can harvest large mite colonies.	Treat transfer points between main and cross belts regularly Remove egg remains from conveyor belts.

Table 2 continued

Hazard category	Hazard	Likelihood	Severity	Risk	Motivation	Corrective action
Eggs	Spread of mites within poultry house via conveyor belts	3	3	9	The likelihood that mites hide on egg conveyor belts is relatively large, especially for woven belts.	Remove egg remains Keep conveyor belts free of dust Do not use woven belts Treat belts with silica dust
Manure aeration	Spread of mites via manure aeration pipes	3	3	9	With intermittent aeration, pipes offer a good hiding place for mites. After a period of non-aeration, mites could be spread throughout the barn if aeration is switched on again.	Use continuous aeration Blow silica dust through the aeration system
Manure aeration	Spread of mites via air mixing box	1	3	3	Mites could hide in ventilation shafts and end up in the air mixing box. This may result in spread of the mites throughout the barn. Likelihood is low, because there is only one air mixing box per barn.	Treat air mixing box with control agent.
Cadavers	Introduction of mites from infested cadaver dump	3	1	3	If the cadaver dump is located alongside the poultry house, mites could enter from here.	Do not place cadaver dump alongside the poultry house Provide a barrier of silica dust between cadaver dump and poultry house
Cadavers	Spread of mites via removal of cadavers	3	3	9	Cadavers that remain in the poultry house for an extended period (>24 h) offer excellent hiding places for mites. During cadaver removal, mites could drop on other surfaces and infest other places in the poultry house.	Remove cadavers the same day Collect cadavers in a clean plastic bag or bucket
Visitors/ external personnel	Introduction of mites via visitors or external personnel	2	2	4	If visitors/external personnel have been in contact with mites, these mites could be transmitted to the next farm. Occasional visits should be limited to only a few places (medium likelihood and severity).	Prohibit visitors from poultry house as much as possible Visitors should shower, wear work clothing and a hairnet

Table 2 continued

Hazard category	Hazard	Likelihood	Severity	Risk	Motivation	Corrective action
Visitors/ external personnel	Spread of mites between barns via visitors or external personnel	2	2	4	Mites may land or crawl onto visitors/ external personnel during their visit. Mites may thus be carried to other barns that are visited.	Prohibit visitors as much as possible Change outer clothing per barn Possibly shower per barn visited
Visitors/ external personnel	Spread of mites within poultry house via visitors or external personnel	3	2	6	Similarly, visitors or external personnel may spread mites within the poultry house. They will probably visit only a limited number of places.	Do not allow visitors.
Poultry farmer/ employee	Introduction of mites via poultry farmer/employee	2	3	6	Farmers and employees could carry mites into the poultry house after contacts outside of their poultry house. Severity is high, because these persons move freely throughout the entire poultry facility	Showering, wearing work clothing and a hair net.
Poultry farmer/ employee	Spread of mites between barns via poultry farmer/employee	3	3	9	Farmer and employees touch hens and the housing system. Thus, there is a high likelihood of infestation and transport of mites to other barns. Severity is high, because these persons move freely throughout the entire poultry facility.	Change outer clothing per barn
Poultry farmer/ employee	Spread of mites within barn via poultry farmer/employee	3	3	9	Similarly, mites may be spread within a barn	Limit movement throughout the barn
Ventilation	Introduction of mites via intake of outside air that is infested with mites	2	2	4	Mites are extremely light (adults ~ 0.075 mg; nymphs ~ 0.01–0.025 mg). Mites that are in the direct surroundings of the poultry house could be sucked in through the ventilation system.	Keep direct surroundings of the poultry house free of bird nests that may harbour mites Do not use pressurized ventilation Air inlet through top of the roof

Table 2 continued

Hazard category	Hazard	Likelihood	Severity	Risk	Motivation	Corrective action
Ventilation	Spread of mites between barns via air ventilation currents	1	2	2	Air that is withdrawn from the barn via the ventilation fans, could be carried to other barns. However, the likelihood of this happening is small.	
Ventilation	Spread of mites within barns via air currents	1	1	1	Air currents within a barn are small and the likelihood for spread via these currents is low.	
Unproductive hens	Introduction of mites via hen collecting teams	1	3	3	Hen collection teams visit several farms and could transfer an infestation. Hen collection takes place only once per round (low likelihood), but personnel visit the entire barn (high severity).	Because hen collection staff visit several farms, they should arrive clean or take a shower Wear clean work clothing
Unproductive hens	Introduction of mites via containers or crates	1	3	3	Containers and crates used for transport of hens visit different farms and could carry an infestation. This only happens once per round (low likelihood), but containers/crates are used throughout the entire barn (high severity).	Containers/crates for transport of hens should be cleaned

hazards or reduce them to an acceptable level. In our study, a point, step or procedure was regarded as a CCP when the calculated risk had a value of 3 or higher on a scale of 1–9. Based on these criteria, 31 of the 41 hazards could be regarded as CCPs. Table 2 shows the results of the hazard analysis and identification of CCPs together with suggestions for control measures (corrective actions) for prevention of *D. gallinae* infestations and spread in poultry facilities.

Establishing critical limits

Establishment of critical limits for each CCP for infestation and spread of *D. gallinae* in poultry facilities is difficult. The logical aim is a critical limit of zero mites, because under optimal conditions, introduction of only a few mites could develop into a major infestation within a few weeks, especially when measures for effectively killing the mites are not available. However, the current literature does not quantify the relationships between mite infestation level and the risk factors present in and outside the poultry facility and poultry welfare/health.

Monitoring

Monitoring of CCPs is a matter of regular and thorough checking of possible entry routes for mites, either in relation to the structure of the poultry facility (barn design and immediate surroundings, ventilation system, etc.) or to those elements that regularly enter or leave the poultry facility (feed, manure, workers, etc.; Table 2). In addition to monitoring the flocks for poultry red mite presence it is of utmost importance that a subsequent quick response is possible to limit an increase of the mite population. Detrimental effects of *D. gallinae* and extra costs of mite eradication can be reduced when early awareness leads to isolation of restricted infected zones rather than a complete layer house (Mul et al. unpublished data). Due to the small size of the mite and vast number in which it aggregates, it is difficult for existing monitoring methods to provide accurate estimates of actual *D. gallinae* numbers (Nordenfors and Chirico 2001). At present, infestations of *D. gallinae* are mostly noticed when farmers or workers are bitten by *D. gallinae*, when mites are seen on the belt and feeders, clumps of mites are seen or when blood spots are detected on eggs. In the Netherlands, farmers were made more aware of *D. gallinae* infestations by using traps consisting of PVC tubing containing a wooden stick as an attractive hiding place for mites (Van Emous, personal communication). These traps were installed throughout the houses of the laying hens to identify the best location. The scale for scoring mite density (score 0–5; no poultry red mites—very many poultry red mites) is quite rough and insensitive to small changes in infestation level. Therefore cases of extreme infestations (higher than "many") remain difficult to quantify. Similar monitoring tools involve corrugated cardboard traps (Nordenfors and Chirico 2001), the ADAS monitoring trap and a trap consisting of a tube containing a fabric or cloth (Maurer et al. 1993). An alternative trap is treated with acaricides (Chirico and Tauson 2001; Lundh et al. 2005). Applying traps in the poultry house alone will not prevent the introduction and spread of *D. gallinae*, but is merely intended to detect infestations and monitor population trends. The farmer needs to be aware that improvements in hygiene (extra cleaning) will reduce the number of mites, but only for a limited period (Maurer, personal communication).

Corrective actions

An overview of possible corrective actions is provided in Table 2. Establishing corrective actions is a continuing process that should be repeated regularly. A farmer should check his farm by going through a checklist (see below) every few months. If the checklist indicates that corrective action is required, then this should be performed immediately in order to limit infestation and spread of *D. gallinae*.

Documentation and validation

From a practical point of view, farmers can plan in advance on a calendar when to go through the checklist. These checklists should be archived and well documented to show whether they are performed regularly and if necessary when and where corrective actions have been carried out. Documentation of date and place of treatment within the poultry facilities provide information concerning the effectiveness of treatments and indicate emergence of resistance to chemical control agents, especially when compared to records from other (nearby) poultry facilities. Validation of the corrective actions should be tested in research or farm trials. Collection of all available farm data on a regional and national basis may prove to be a valuable tool in the evaluation of corrective actions.

Checklist

As an extra management aid to farmers, we have prepared a checklist to help identify the most important points of action in the prevention of *D. gallinae* infestations and spread. This checklist was evaluated during an in depth workshop by five Dutch poultry farmers and briefly by 40 British poultry farmers during a course on *D. gallinae*. The five Dutch farmers identified the checklists added value and improved it. They owned family farms, their laying hens were housed in Dutch barn systems, in free range systems and in cage layer systems with between 20,000 and 100,000 birds. The Dutch farmers described their current prevention measures with regard to infestation and spread of *D. gallinae* at their poultry farm. Before providing the checklist, farmers were encouraged to discuss their measures. During the discussions, several preventive actions were suggested and added to the farmers own lists of preventive measures. Their suggestions included:

- Heating the henhouse to temperatures above 55°C.
- Regular washing down of the housing system.
- Treatment of the walls and floors with silica dust or carbolineum prior to introduction of the new hens.

Subsequently, the farmers received the checklist and were asked to fill it out to ensure its feasibility and usefulness and to indicate which measures were additional to their list of preventive measures made prior to receiving the checklist. The farmers' advice led to a new draft in which questions were removed and/or adapted. All five farmers indicated that the checklist had encouraged them to take new or alternative preventive measures including:

- Checking if all persons and material entering the farm were free of *D. gallinae*.
- Placement of cobblestones directly around the poultry facilities in order to reduce the number of pests that are potential carriers of mites.
- Treating the edges of the feeding troughs with silica dust or glue.

- Treatment of the manure conveyor belt.
- Checking the cleanliness of egg trays.
- Treating the air mixing box.
- Order *D. gallinae* free growing hens.
- Monitoring poultry facilities for *D. gallinae*.

The overall conclusion of the five farmers was that the checklist was potentially a useful tool.

In the UK, the checklist was adapted to the egg production system in the UK. Of the 40 British poultry farmers, 28 own battery units with between 40,000 and 100,000 birds, 8 poultry farmers own free-range units with between 3,000 and 8,000 birds and 4 poultry farmers produce eggs in barns with houses for between 5,000 and 8,000 birds. The most interesting remarks of the egg producers were that (1) UK egg producers never have hobby birds at the site, (2) corrugated roofs are always insulated, (3) only very few farms have a shower, (4) cadaver dumps are not used in the UK, (5) workers often wear the same overalls all week and do not change between units, and (6) the use of silica dust on conveyor belts was thought to be a useful recommendation. They mentioned that the checklist stimulated them to be more critical about the way they run their units and highlighted things that could be improved. All found the checklist feasible and useful. The adjusted and final checklist can be obtained through the first author.

Conclusions and discussion

Because poultry red mite is considered a major hazard to the health and welfare of poultry (Hegelund and Sørensen 2007), we elaborated on all possible risk factors for introduction and spread of *D. gallinae* by conducting an analysis using the HACCP method. In general, this method was evaluated as very helpful. We should note that the checklist is based on opinions of a limited number of experts and poultry farmers who were involved in this study. For example, it is possible that other experts would assign different likelihood and severity scores because of differing conditions in other regions or countries (e.g., differences in housing systems or environment in the vicinity of the poultry facility) or other judgements. Therefore, we consider it desirable to evaluate the developed procedures for other (European) countries. However, a farmer will only use the developed checklist when he is aware of the variety of effects of a *D. gallinae* infestation. Therefore, we argue that educative measures should be undertaken. For example, educative illustrations that show the possible points of introduction and routes of further spread of *D. gallinae* may be a good way to spread information. A major challenge remains in determination of critical limits of control points in unique situations, and, when and how to take action when such limits are exceeded. More quantitative, epidemiological studies are essential for the provision of clear targets for effective on-farm mite control. Additionally, more studies are needed that elucidate the various behaviors of *D. gallinae*, such as host seeking and aggregation. Insights obtained from such studies could be used to improve understanding of the various routes of infestation and spread. Moreover, this information will facilitate the development of alternative and environmentally safe control methods such as those based on the attract and kill principle (Stetter and Lieb 2000). Finally, although the HACCP process may seem cumbersome and extensive, we agree with an earlier statement that: "the concept is structuring and formalizing what truly good farmers are doing anyway" (after Ryan 1997, in: Noordhuizen and Frankena 1999).

Acknowledgments This research was funded by the Dutch Product Board for Poultry and Eggs and the Dutch Ministry of Agriculture, Nature and Food Quality. We are grateful to T. Fiks-Van Niekerk, R. van Emous and B. Reuvekamp (Animal Sciences Group, Wageningen University) for their knowledgeable contribution to this research. We would like to thank all Dutch and British farmers that participated in the evaluation of the checklist. We also thank B. Bell (ADAS) for testing the checklist during her course for poultry farmers on *D. gallinae*.

References

Baselga E, Drolet BA, Esterly NB (1996) Avian mite dermatitis. Pediatrics 97(5):743–745

Bonde MJ, Sørensen JT (2004) Herd health management in organic pig production using a quality assurance system based on hazard analysis and critical control Points. NJAS-Wag J Life Sci 52:133–143

Chauve CM (1998) The poultry red mite *Dermanyssus gallinae* (De Geer, 1778) : current situation and future prospects for control. Vet Parasitol 79:239–245. doi:10.1016/S0304-4017(98)00167-8

Chirico J, Tauson R (2001) Traps containing acaricides for the control of *Dermanyssus gallinae*. Vet Parasitol 110:109–116. doi:10.1016/S0304-4017(02)00310-2

Hegelund L, Sørensen JT (2007) Developing a HACCP-like system for improving animal health and welfare in organic egg production—based on an expert panel analysis. Animal 1:1018–1025. doi:10.1017/S1751731107000213

Kijlstra A, Meerburg BG, Mul M (2004) Animal-friendly production systems may cause re-emergence of *Toxoplasma gondii*. NJAS-Wag J Life Sci 52(2):119–132

Kilpinen O, Roepstorff A, Permin A, Nørgaard-Nielsen G, Lawson LG, Simonsen HB (2005) Influence of *Dermanyssus gallinae* and *Ascaridia galli* infections on behaviour and health of lying hens (*Gallus gallus domesticus*). Br Poult Sci 45(1):26–34. doi:10.1080/00071660400023839

Lundh J, Wiktelius D, Chirico J (2005) Azadirachtin-impregnated traps for the control of *Dermanyssus gallinae*. Vet Parasitol 130:337–342. doi:10.1016/j.vetpar.2005.02.012

Marangi M, Giangaspero A, Raele D, Assunta Cafiero M, Camarda A, Sparagano OAE (2008). Evaluation of poultry red mite susceptibility, *Dermanyssus gallinae* (Acarina: Dermanyssidae), to some acaricides in Italy. In: Abstracts of the BSP Spring, Trypanosomiasis/Leishmaniasis & Malaria Meetings. Newcastle upon Tyne, UK 168, 30 March, 2 April

Maurer V, Hertzberg H (2001) Ökologische Legehennenhaltung. Was tun die kleinen Vampire? DGS-Mag 40:49–52

Maurer V, Perler E (2006). Silicas for control of the poultry red mite *Dermanyssus gallinae*. Paper presented at the Joint Organic Congress, Odense, Denmark, 30–31 May (http://orgprints.org/7274/01/Maurer_Dermanyssus_Odense.pdf)

Maurer V, Baumgärtner J, Bieri M, Fölsch DW (1993) The occurrence of the chicken mite *Dermanyssus gallinae* in Swiss poultry houses. Mitt Schweiz Entomol Ges 66:87–97

Mayes T (1993) The application of management systems to food safety and quality. Trends Food Sci Technol 4:216–219. doi:10.1016/0924-2244(93)90154-3

Mayes T (1998) Risk analysis in HACCP: burden or benefit? Food Contr 9:171–176. doi:10.1016/S0956-7135(97)00055-8

Moro CV, Chauve C, Zenner L (2005) Vectorial role of some Dermanyssoid mites (Acari, Mesostigmata, Dermanyssoidea). Parasite 12:99–109

Mouchtouri VA, Anagnostopoulou R, Samanidou-Voyadjoglou A, Theodoridou K, Hatzoglou C, Kremastinou J, Hadjichristodoulou C (2008) Surveillance study of vector species on board passenger ships risk factors related to infestations. BMC Public Health 8:100. doi:10.1186/1471-2458-8-100

Mul M, Van der Gaag M (2003). Control of Salmonella at pig finishing farms with a farm decision tree. Proceedings of the 5th International Symposium on the Epidemiology and Control of Foodborn Pathogens in Pork, Crete, Greece, 149–151

Noordhuizen JPTM, Frankena K (1999) Epidemiology and quality assurance: applications at farm level. Prev Vet Med 39:93–110. doi:10.1016/S0167-5877(98)00151-2

Nordenfors H (2000). Epidemiology and control of the Poultry Red Mite, *Dermanyssus gallinae*. Doctoral thesis, Swedish University of Agricultural Sciences, Uppsala, Sweden, 19–20

Nordenfors H, Chirico J (2001) Evaluation of a sampling trap for *Dermanyssus gallinae* (Acari: Dermanyssidae). J Econ Entomol 94:1617–1621

Nordenfors H, Höglund J, Tauson R, Chirico J (2001) Effect of permethrin impregnated plastic strips on *Dermanyssus gallinae* in loose-housing systems for laying hens. Vet Parasitol 102:121–131. doi:10.1016/S0304-4017(01)00528-3

Rayment GE (2006) Australian efforts to prevent the accidental movement of pests and diseases in soil and plant samples. Commun Soil Sci Plant Anal 37:2107–2117. doi:10.1080/00103620600817069

Sikes RK, Chamberlain RW (1954) Laboratory observations on three species of bird mites. J Parasitol 40:691–697. doi:10.2307/3273713

Steenberg T, Kilpinen O, Jespersen JB, Soler-Cruz MD, Vega-Robles C, Birkett M, Dewhirst S, Pickett J (2005) Advances in the development of novel control methods against chicken mites (*Dermanyssus gallinae*). Bull-OILB/SROP 28:89–91

Stetter J, Lieb F (2000) Innovation in crop protection: trends in research. Angew Chem Int Ed 39:1724–1744. doi:10.1002/(SICI)1521-3773(20000515)39:10<;1724::AID-ANIE1724>;3.0.CO;2-5

Sun YM, Ockerman HW (2005) A review of the needs and current applications of hazard analysis and critical control point (HACCP) system in foodservice areas. Food Contr 16:325–332. doi:10.1016/j.foodcont.2004.03.012

Wood HP (1917). The chicken mite: its life history and habits. USDA Bulletin No. 553